STRUCTURE
AND BONDING

Volume 29

Editors: J. D. Dunitz, Zürich
P. Hemmerich, Konstanz · J. A. Ibers, Evanston
C. K. Jørgensen, Genève · J. B. Neilands, Berkeley
D. Reinen, Marburg · R. J. P. Williams, Oxford

With 51 Figures and 48 Tables

Springer-Verlag Berlin Heidelberg GmbH

ISBN 978-3-662-15494-6 ISBN 978-3-540-38055-9 (eBook)
DOI 10.1007/978-3-540-38055-9

Library of Congress Catalog Card Number 67-11280

© Springer-Verlag Berlin Heidelberg 1976
Originally published by Springer-Verlag Berlin Heidelberg New York 1976
Softcover reprint of the hardcover 1st edition 1976

Typesetting: R. & J. Blank, München.

Contents

STRUCTURE AND BONDING is issued at irregular intervals, according to the material received. With the acceptance for publication of a manuscript, copyright of all countries is vested exclusively in the publisher. Only papers not previously published elsewhere should be submitted. Likewise, the author guarantees against subsequent publication elsewhere. The text should be as clear and concise as possible, the manuscript written on one side of the paper only. Illustrations should be limited to those actually necessary.

Manuscripts will be accepted by the editors:

Professor Dr. *Jack D. Dunitz*   Laboratorium für Organische Chemie der Eidgenössischen Hochschule
CH-8006 Zürich, Universitätsstraße 6/8

Professor Dr. *Peter Hemmerich*   Universität Konstanz, Fachbereich Biologie
D-7750 Konstanz, Postfach 733

Professor *James A. Ibers*   Department of Chemistry, Northwestern University
Evanston, Illinois 60201/U.S.A.

Professor Dr. *C. Klixbüll Jørgensen*   51, Route de Frontenex,
CH-1207 Genève

Professor *Joe B. Neilands*   University of California, Biochemistry Department
Berkeley, California 94720/U.S.A.

Professor Dr. *Dirk Reinen*   Fachbereich Chemie der Universität Marburg
D-3550 Marburg, Gutenbergstraße 18

Professor *Robert Joseph P. Williams*   Wadham College, Inorganic Chemistry Laboratory
Oxford OX1 3QR/Great Britain

SPRINGER-VERLAG

D-6900 Heidelberg 1
P. O. Box 105280
Telephone (06221) 4 87·1
Telex 04-61723

D-1000 Berlin 33
Heidelberger Platz 3
Telephone (030) 82 20 01
Telex 01-83319

SPRINGER-VERLAG
NEW YORK INC.

175, Fifth Avenue
New York, N. Y. 10010
Telephone 673-2660

The Molecular Basis of Biological Dinitrogen Fixation

Walter G. Zumft

Fachbereich Biologie-Chemie der Universität, 8520 Erlangen, Germany.

Table of Contents

Abbreviations

ATP	= Adenosine 5'-triphosphate;		K_d	= Dissociation constant;
ADP	= Adenosine 5'-diphosphate;		*his*	= Histidine operon;
CTP	= Cytidine 5'-triphosphate;		*his* D	= Histidinol dehydrogenase;
GTP	= Guanosine 5'-triphosphate;		*gnd*	= Gluconate-6-phosphate dehydrogenase;
UTP	= Uridine 5'-triphosphate;			
NAD(H)	= Nicotinamide adenine dinucleotide or its reduced form;		*nif*	= Operon(s) for dinitrogen fixation;
NADP(H)	= Nicotinamide adenine dinucleotide phosphate or its reduced form;		*rfb*	= Gene cluster for cell-wall lipopolysaccharide synthesis (related to phage resistance);
CCCP	= *m*-Cl-carbonyl cyanide phenylhydrazone;		*shu*	= Shikimate utilization;
			Fe protein	= Iron protein (the low-molecular-weight component of nitrogenase);
EPR	= Electron paramagnetic resonance;			
NMR	= Nuclear magnetic resonance;		MoFe protein	= Molybdenum-iron protein (the high-molecular-weight component of nitrogenase);
FMN	= Flavin mononucleotide;			
Ph	= Phenyl;		$E_{m\,7.5}$	= Mid-point potential at pH 7,5;
Cp	= Cyclopentadienyl;		Mg · ATP	= Simplified notation for the two-fold negatively charged magnesium-ATP complex at physiological pH;
Et	= Ethyl;			
K_m	= Michaelis constant;			
K_i	= Inhibition constant;			

2

I. Introduction

Three to four billion years ago, when the primordial earth was rich in highly reactive compounds containing triple bonds, such as cyanide, cyanogen, nitriles, isonitriles and others, it is speculated that an enzyme may have evolved which reduced and detoxified these compounds to ensure the survival of primeval organisms (1). Today, the same enzyme catalyses the reduction of molecular nitrogen to ammonia and constitutes a prime link in the nitrogen cycle of the earth. Dinitrogen fixation, either as the biological process or as its industrial variant, the Haber-Bosch ammonia synthesis, is the limiting factor in supplying a reduced form of nitrogen to bacteria and plants, thus directly affecting protein synthesis. A practically unlimited reservoir of dinitrogen exists in the atmosphere of the earth. To take advantage of this source, mankind must use its limited and costly sources of energy. In a steadily growing world with increasing demands of fixed nitrogen and finite energy sources, imitating the biological process of homogeneous catalysis under mild reaction conditions would help to meet this demand. Two ways to achieve dinitrogen fixation under mild conditions are being investigated: first, the biochemical analysis of the biological process, and second, the synthesis and analysis of transition-metal complexes which are capable of dinitrogen reduction in protic media. Historically, the first investigations did not begin with the preparation of N_2-fixing cell-free extracts in 1960, since many data and characteristics of the enzyme system had been deduced earlier from studies *in vivo* (2, 3). The long sought after *in vitro* systems presented a kinetic barrier to further studies and once it was resolved research in this area accelerated tremendously. The chemical approach experienced a similar event with the first synthesis of a dinitrogen complex in 1965 by *Allen* and *Senoff* (4). Since then, major advances have been made, albeit the ultimate breakthrough in both fields appears to be still before us.

Many concepts and areas investigated in various laboratories at present, are based on work done during the past decade and cannot be easily described when separated from it. A broader coverage is therefore attempted to account for this situation; nevertheless, there had to be restrictions. They were applied in rather strict manner outside the biochemical events that transform molecular nitrogen to ammonia. Biological problems in a broader sense, and the metabolic fate of the NH_3 molecule are not discussed. The chemical aspects of dinitrogen fixation have been covered in recent reviews, such as the chemistry and reactivity of dinitrogen (5), or the synthesis, properties, and reactivity of dinitrogen complexes of the transition metals (6–9). Specific inorganic models for the biological process are treated by *Leigh* (5), and *Chatt* and *Richards* (6), whereas an article by *Hardy et al.* (7) aims at an integrated presentation of biological and inorganic dinitrogen fixation.

Nomenclature. Burk, in 1934, named the terminal dinitrogen reducing complex *nitrogenase* (10). No systematic name has been selected for this enzyme, although *Burris* (11) proposed *ferredoxin* : N_2 *oxidoreductase* (ATP hydrolyzing). Nitric oxide reductase [EC 1.7.99.2], a flavoprotein which reduces nitric oxide to dinitrogen is sometimes (and admittedly correctly) termed "nitrogenase", yet has no relation to the enzyme described here.

Nitrogenase acts as a complex of two different proteins which will be called iron protein (Fe protein) and molybdenum-iron protein (MoFe protein). This nomenclature is actually widely accepted and is preferentially used in a monograph on the biochemistry of dinitrogen fixation (12). Equivalent names for the MoFe protein found in the literature are component I, molybdoferredoxin, azofermo or Kp 1; the Fe protein is occasionally referred to as component II, azoferredoxin, azofer, or Kp 2. According to the International Union of Pure and Applied Chemistry (Nomenclature of Inorganic Chemistry, 1957) the nitrogen molecule, N_2, will be called dinitrogen; the oxygen and hydrogen molecules dioxygen and dihydrogen, respectively.

II. Reaction Requirements

Nitrogenase is a multimeric protein complex which contains molybdenum and a large number of iron-sulfur groups. It is now firmly established that the enzyme catalyzes a reductive breakdown of the nitrogen molecule with ammonia as the sole reaction product. ATP is obligatorily hydrolyzed during the reaction to ADP and P_i; neither AMP nor pyrophosphate have been detected. In addition to dinitrogen reduction and ATP hydrolysis, the enzyme catalyzes the evolution of dihydrogen from hydronium ions. This process is competitive with dinitrogen reduction.

Nitrogenase consists of two different proteins which must act concertedly for catalysis of any of the above-mentioned reactions. No partial reactivity has been found for the separated components from any biological source. Ferredoxin and flavodoxin are the physiological electron donors for clostridial and other nitrogenases. Most *in vitro* assay systems, however, use sodium dithionite as electron donor to avoid the necessity of coupling the enzyme to its physiological reductant source (in many cases the latter has not yet been established). The ATP requirement of the reaction is satisfied by substrate amounts of the nucleotide, or, since ADP is a powerful inhibitor of dinitrogen reduction, by the use of an ATP-generating system. Creatine phosphate/creatine kinase, or acetyl phosphate/acetate kinase are commonly used for this purpose. Divalent metal ions are a further requisite for the nitrogenase reaction even in the absence of a phosphorylating system. This is due to the necessity of complex formation of ATP with a metal before interaction with the enzyme. Magnesium, the most effective metal in the nitrogenase reaction, can be replaced by Mn^{2+}, Co^{2+}, Fe^{2+} and Ni^{2+} in this order of decreasing reactivity. Copper and Zn^{2+} are inhibitory. Nitrogenase shows an exceptional versatility with substrates. In addition to dinitrogen, other compounds with triple bonds such as N_3^-, N_2O, C_2H_2, HCN, nitriles, and isonitriles are reduced. Their reduction shows the same general requirements as dinitrogen reduction.

Acetylene reduction is widely used to measure nitrogenase activity because of the ease of quantitation of its reaction product, ethylene, by gas chromatography (13). Dinitrogen reduction itself can be followed by the ^{15}N isotope technique (14) or by microdiffusion of the ammonia formed and subsequent colorimetric estimation (*15, 16, 17*). Dihydrogen evolution is conveniently followed by manometric methods (*18*). ATP hydrolysis can be monitored by following P_i release (*19, 20*) or a reaction product of a generating system, for instance, creatine (*21*). The enzyme has an activity optimum between pH 7 and 8 and demands strictly anaerobic conditions for study *in vitro*. Figure 1 shows a generalized scheme of the nitrogenase reaction.

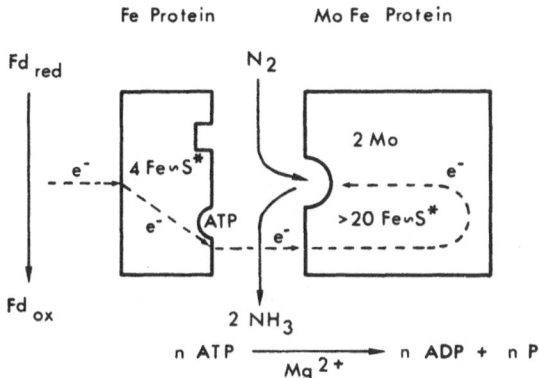

Fig. 1. The basic requirements of reduction of dinitrogen to ammonia by nitrogenase. Ferredoxin (Fd) serves as the reductant in this scheme. Electrons are initially transferred to the Fe protein, the low-molecular-weight part of the enzyme. This protein has two binding sites for $MgATP^{2-}$ and undergoes a conformational change upon nucleotide binding. Electron transfer from the Fe protein to the MoFe protein, the high-molecular-weight part of the enzyme, is ATP dependent. Both parts of the enzyme contain numerous iron-sulfur groups which participate in the electron transfer, and may also serve, especially in the MoFe protein, as electron storage units for the active site.

III. Sources and Preparation of Nitrogenase

A. Biological Distribution

The ability to fix dinitrogen appears to be restricted to prokaryotic organisms, *i.e.* bacteria and blue-green algae. Though this cannot be stated categorically, no confirmed report exists of dinitrogen fixation in eukaryotes. It has been claimed repeatedly that some fungi have the ability to fix dinitrogen, but investigation of several doubtful species with $^{15}N_2$ or the acetylene assay did not substantiate these claims (*22*). Dinitrogen fixation is common among primitive organisms like sulfate-reducing bacteria, photosynthetic sulfur and nonsulfur bacteria, and methanogenic and hydrogen bacteria. The reaction is mostly associated with an anaerobic type of metabolism. Anaerobic conditions are required for several facultative anaerobic bacteria to fix dinitrogen and only a few organisms are capable of using dinitrogen in a completely aerobic environment. Most, if not all photosynthetic bacteria will reduce N_2. The reaction is also common among blue-green algae. Many of these algae possess specialized cells, so-called heterocysts, whose function and involvement in dinitrogen fixation continues to be a hotly debated topic. The reader is referred to an extensive review by *Stewart* (23) concerning all aspects of dinitrogen fixation by photosynthetic organisms. A few representatives of each type of dinitrogen fixing organisms are listed in Table 1.

Table 1. Representatives of N_2-fixing organisms

Type of organism	Species	Ref.
BACTERIA		
Anaerobic	*Clostridium pasteurianum*	(24)
	Desulfovibrio vulgaris	(331)
	Desulfotomaculum ruminis	(332)
Facultative anaerobic	*Klebsiella pneumoniae*	(333)
	Bacillus polymyxa	(334)
Aerobic	*Azotobacter vinelandii*	(174)
Photosynthetic	*Rhodospirillum rubrum*	(120)
Symbiotic	*Rhizobium japonicum*	(335)
BLUE-GREEN ALGAE		
Free-living	*Anabaena cylindrica*	(336)
	Gloeocapsa alpicola	(337)
Symbiotic	*Anabaena azollae*	(338)

B. Preparation of Nitrogenase

It is the studies by *Carnahan et al.* (24) on the efficiency of various substrates to support dinitrogen fixation that became crucial in obtaining cell-free systems with a reproducible and appreciable capacity to reduce dinitrogen. Nitrogenase activity was found prior to 1960 in "cell-free extracts" of bacteria and blue-green algae, but since various requirements of the nitrogenase reaction were unknown at that time, this work was difficult to reproduce and had little impact upon the elucidation of the enzymatic mechanism. Pyruvate proved to be the best substrate for maintaining dinitrogen fixation in clostridial extracts and although *Carnahan et al.* (24) presumed that pyruvate supplied reducing equivalents, the mode of coupling pyruvate oxidation to dinitrogen fixation remained obscure. While this relation was being investigated, a new electron transfer protein, ferredoxin, was isolated and was shown to be the immediate electron donor for nitrogenase (25). The choice of pyruvate in this early enzymatic work on nitrogenase was fortunate, since ATP is formed simultaneously during pyruvate breakdown. An ATP requirement of nitrogenase was not recognized until it was inferred from inhibition by glucokinase and arsenate that ATP may contribute directly to N_2 fixation (26). Previously, ATP had even been found to be inhibitory (24). The dual requirement of reductant and ATP may account for many of the earlier failures to obtain N_2 fixation *in vitro*.

Cold lability, oxygen sensitivity, and a narrow stability range of the N_2-fixing system towards the hydrogen-ion concentration were other factors which prevented the facile isolation of the enzyme system from an organism. Now that most critical factors are known, the purification of nitrogenase is relatively easy and comprises only a few fractionation steps. Preparation of nitrogenase is aided by the large amount of this enzyme in certain organisms, where under proper conditions the enzyme can comprise several percent of the total cell protein.

Nitrogenase has been isolated from about twenty species of bacteria and blue-green algae. In fewer cases a separation into two components has been achieved: *Azotobacter vinelandii (27)*, *A. chroococcum (28)*, *Bacillus polymyxa (29)*, *Clostridium pasteurianum (30)*, *Klebsiella pneumoniae (29)*, *Mycobacterium flavum (31)*, *Rhizobium japonicum (32)*, *R. lupini (33)*, *Anabaena cylindrica (34)*, and *Chromatium* Strain D *(35)*.

No general purification scheme is applicable to all nitrogenases, although only a few separation principles are in common use. When it became evident that nitrogenase consists of two proteins, purification became concentrated on these single components rather than on the nitrogenase complex. However, the preparation of an active nitrogenase complex has been preferred for kinetic *(36)* and stoichiometric studies *(37, 38)* rather than the use of an arbitrarily reconstituted enzyme.

The nitrogenase components are frequently obtained by fractionated precipitation with protamine sulfate, polyethylene glycol, or other protein-precipitating agents. This technique is combined with ion exchange and gel chromatography, methods which take advantage of the different acidity and the difference in molecular weight of the two components.

Homogeneous MoFe proteins and Fe proteins have been obtained from *A. vinelandii (39, 40, 41)*, *C. pasteurianum (42, 43)*, *K. pneumoniae*, *(44)* and two *Rhizobium* species *(33, 45)*. The MoFe protein from *A. vinelandii* has been crystallized in the form of white or brown needles with dimensions of $25-60\ \mu m$ to $1-4\ \mu m$ *(39, 40)*.

The entire preparation of the two nitrogenase proteins must be performed under anaerobic conditions because both proteins are rapidly inactivated by air. Fortunately, the purified components will retain their activity in liquid nitrogen over prolonged storage periods *(46)*. Also, the unfractionated nitrogenase complex from *Azotobacter* is less oxygen-sensitive than the separated and purified components.

C. Inactive Species

The protein purity of both nitrogenase components is now proven by several criteria, but homogeneity with respect to the catalytic center(s) still remains problematic. Several lines of evidence point to an active-site heterogeneity of otherwise pure components. The specific activity of highly purified proteins varies by a factor of 3 to 4. Acetylene-reducing activity of the MoFe proteins has been reported as 704 to 933 units for *Rhizobium (33, 45)*, 1200 to 2000 units for *Klebsiella (44, 47)*, 1400 to 1638 units for crystalline preparations from *Azotobacter (40, 48)*, and 1120 to 2750 units for the clostridial MoFe protein *(42, 49)*. The Fe proteins show similarly wide ranges of activity: 980 to 1100 units for the

Klebsiella (*44, 47*), 2708 to 3100 units for the clostridial (*50, 51*), and 2128 units for the *Azotobacter* protein (*52*). Part of this variation may be due to *art* specific differences since not all nitrogenases ought to have the same turnover number, yet, considerable variation has been noted among laboratories working with the same or closely related organisms or even among experimenters working in the same laboratory.

The molybdenum content is variable in MoFe proteins from different sources, which appears inconceivable on the grounds of a generally assumed key role of molybdenum for catalysis. Considerable differences also apply to the iron content of different MoFe proteins.

Double-integrated signal intensities of the EPR spectrum of different Fe proteins range from 0.2 to 1 mole electron/mole protein (*53–56*); likewise, the clostridial MoFe protein yields only a fraction of one electron per protein molecule (*53*). In addition to these variable and low integration values, EPR spectroscopy has detected a signal around $g = 1.94$ in MoFe proteins (*35, 44, 48, 53*) which showed a diverging temperature dependence from the low-field signals (*35, 53*) and finally disappeared from further purified samples (Fig. 2, for discussion of the EPR spectra see Section IV. C). Heterogeneity of the MoFe proteins

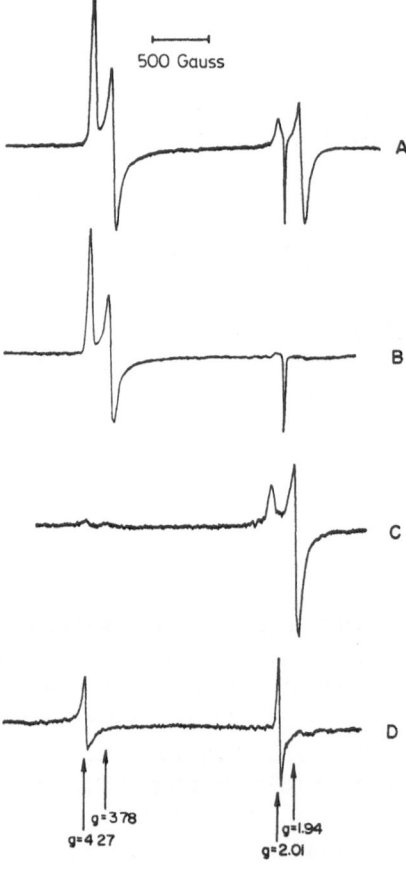

Fig. 2. EPR spectra of the MoFe protein and its demolybdo form. (A) Spectrum of unfractionated MoFe protein; protein concn. 72 mg/ml; specific activity 1200 units/mg; note the distinct resonance at $g = 1.94$ which accounts for about 40% contamination of the MoFe protein with the demolybdo form. (B) Spectrum of highly purified MoFe protein; protein concn. 36 mg/ml; specific activity 2720 units/mg; g-values, 4.27, 3.78, and 2.01. (C) Spectrum of the demolybdo form of the MoFe protein; protein concn. 20 mg/ml; g-values, 2.05 and 1.92; this protein is catalytically inactive when combined with the Fe protein. (D) like (C) after exposure for 5 min to air. Spectrometer settings: sweep rate, 500 gauss/min; time constant, 0.3 s; frequency, 9.22 GHz; modulation amplitude, 4.5 Gauss; gain 50 (A, B) or 100 (C, D); temperature, 23 *K*. Reproduced by permission of the Fed. of Eur. Biochem. Soc. from Ref. (*42*).

of *Klebsiella* (*44*) and *Azotobacter* (*41*) is indicated by the formation of two bands during isoelectric focusing.

Most of these observations can be satisfactorily explained by assuming the presence of one or possibly several species of either MoFe protein of Fe protein which differ in their metal centers. Indeed, clostridial MoFe protein, which was pure according to several criteria of protein chemistry, could be fractionated by ion-exchange chromatography into a catalytically active form with 2 Mo atoms and approximately 24 iron-sulfur groups and a catalytically inactive form, containing only traces of Mo and lacking two-thirds of the iron-sulfur groups (*42*). Apart from the difference in metal content, both proteins were indistinguishable in their subunit composition, sedimentation behavior, and immunological reaction and were very similar in their amino-acid composition.

The origin of the inactive species of the MoFe protein cannot be satisfactorily explained at present. Though its nature as a preparation artifact might be obvious, its presence has been observed in differently treated material of varying provenance. It has been suggested that this material may be involved in nitrogenase biosynthesis (*42*). In this context, it seems of interest that during initial phases of repression of nitrogenase by ammonia, with *de novo* protein synthesis inhibited by chloramphenicol, nitrogenase formation was observed over a short period (*57*). Reactivation of a demolybdo form of nitrogenase could occur under these conditions, albeit, a more complex system must be involved to supply also the necessary Fe protein for enzymatic activity.

Active-site heterogeneity might still be present in many nitrogenase preparations and some data should be judged in the light of possible heterogeneity. Chances to obtain active-site heterogeneities are not surprising, due to the oxygen, cold, and salt sensitivity of nitrogenase.

IV. Chemical and Physical Properties

A. Chemical Composition

The chemical composition of nitrogenase components from various sources has been investigated repeatedly with respect to metal content, inorganic sulfide, sulfhydryl groups, and amino acids. Some of the earlier data are likely to deviate from the real values inasmuch as the components used were not sufficiently pure for chemical analysis. With further purified proteins now available, the data are still bound to vary depending on the sucessful removal of inactive species from the protein prior to analysis.

Molybdenum and iron are the metal constituents of all nitrogenases. Crystalline MoFe protein from *A. vinelandii* contains 2 Mo, 34 to 38 Fe, and 26 to 28 S^{2-} per molecular weight 270000 (*39*). These values are likely to change, since this preparation has been shown to contain a cytochrome (*40*) and an inactive species (*42*). A reevaluation of highly purified MoFe protein of this organism yielded 1.5 Mo, 24 Fe and 20 S^{2-} per molecular weight 216000 (*41*). According to recent analyses, clostridial MoFe protein contains 2 Mo, 22 to 24 Fe, and about the same amount of inorganic sulfide per molecular weight 220000 (*42, 58*). Although the molar extinction coefficient for the molybdenum-dithiol complex as reported by *Huang* (*59*) and *Huang et al.* (*58*) is grossly incorrect (*60*), the molybdenum content of the clostridial MoFe protein will not vary because the determinations were based on internal standards and were not calculated from the extinction coefficient. The MoFe protein from *K. pneumoniae* contains 1 Mo, 17 to 18 Fe, and 17 S^{2-} per molecular weight 218000 (*44*). Its iron content is almost twice that of an earlier determination (*61*) and has recently been found to be approximately 30 atoms of Fe per protein molecule (*62*). At the same time the molybdenum content increased to 2 atoms of Mo. The MoFe protein from *R. japonicum* has 1.3 g atoms of Mo, 28 to 29 Fe, and 26 S^{2-} per 200000 daltons (*45*), the same protein from *R. lupini*, however, gives slightly lower numbers (*33*). Because of the low molybdenum content, the presence of an inactive species has been suggested (*45*).

Although it is not proven in all cases, the MoFe protein of nitrogenases most probably contains two atoms of molybdenum per protein molecule. *Bray* and *Swann* (*63*) pointed out the frequent occurrence of two atoms of Mo in molybdenum-containing enzymes. Nitrogenase appears to fit this generalization, although due to the poor information concerning the role of molybdenum in many enzymes no mechanistic implication can be derived.

In addition to molybdenum and iron, a high calcium and magnesium content has been found in the MoFe protein from *C. pasteurianum* (*64*) and *K. pneumoniae* (*44*). No specific role can be assigned to either Ca or Mg, but it is noteworthy that Ca has been shown to be involved in dinitrogen fixation by nutritional tests (*2*), *i.e.* by the same technique that implicated Mo and Fe in N_2 fixation long before chemical analysis revealed their presence in the enzyme. In addition to the above-mentioned metals, the MoFe protein of *K. pneumoniae* contains, approximately 1 Zn and 1 Cu per protein molecule (*44*), the role of which is obscure. Zinc has also been found in a third (and therefore atypical) component from *A. vinelandii* (*65*).

The purified Fe proteins from *C. pasteurianum* (*43*) and *K. pneumoniae* (*44*) contain both 4 Fe and 4 S^{2-} per protein molecule; no other metal constituents are present. *Azotobacter* and rhizobial Fe proteins both have an iron content slightly higher than 3 (*33, 41*).

The amino-acid composition of five MoFe proteins has been determined (see Table 2). With the exception of tryptophan which is absent from the Fe proteins, all 18 amino acids are found in the MoFe proteins as well as in the Fe proteins (see Table 3). The acidic amino acids are preponderant in both nitrogenase components. A diverging low tryptophan content of the clostridial MoFe protein has been thoroughly investigated by titration with N-bromosuccinimide and several colorimetric methods and appears to be a main distinction of this protein from other MoFe proteins (*66*, see below). The main features revealed by a statistical treatment of the amino-acid composition are:

i) that MoFe proteins from different organisms are closer related to each other than to their homologous Fe proteins, and

ii) that both proteins show no relation to bacterial ferredoxins (*67*).

Table 2. Amino-acid composition of MoFe proteins from different organisms

	A. vinelandii mol. wt 216000 (41)	C. pasteurianum mol. wt 220000 (66)	K. pneumoniae mol. wt 218000 (44)	R. japonicum mol. wt 200000 (45)	R. lupini mol. wt 200000 (33)
Asx	199	192	210	182	178
Thr	93	120	104	89	93
Ser	92	92	110	109	99
Glx	211	196	206	179	168
Pro	88	86	92	89	77
Gly	163	182	152	170	156
Ala	120	140	158	162	137
Cys	26	40	38	23	30
Val	129	148	124	126	105
Met	61	60	76	39	50
Ile	108	160	100	112	107
Leu	143	138	182	136	124
Tyr	70	84	68	65	73
Phe	99	68	96	80	85
Lys	130	160	102	122	119
His	50	46	48	53	67
Arg	86	62	102	94	89
Trp	25	6	56	29	n.d. [a]

[a]) not determined.

Table 3. Amino-acid composition of Fe proteins from different organisms

	A. vinelandii mol. wt 64000 (41)	C. pasteurianum mol. wt 55000 (66)	K. pneumoniae mol. wt 678000 (44)	R. lupini mol. wt 65000 (33)
Asx	63	45	60	64
Thr	26	24	36	23
Ser	20	24	24	35
Glx	88	70	84	72
Pro	18	18	18	19
Gly	54	62	60	59
Ala	55	40	60	63
Cys	4	12	18	10
Val	51	34	42	36
Met	21	16	36	14
Ile	38	34	48	45
Leu	50	52	42	57
Tyr	15	18	18	22
Phe	15	11	12	85
Lys	22	32	36	30
His	7	4	6	14
Arg	27	24	24	29
Trp	0	0	0	n.d. [a]

[a]) not determined.

It is noteworthy in this context that nitrogenase can be hybridized from components of different dinitrogen-fixing organisms (*29*). These cross reactions, however, are not very specific. The MoFe protein from *K. pneumoniae* forms an enzymatically active nitrogenase complex with the Fe proteins from *A. vinelandii*, *A. chroococcum*, *B. polymyxa*, *M. flavum*, and *R. rubrum* but forms an inactive adduct with the clostridial Fe protein. Likewise, the Fe protein from *A. vinelandii* reacts with the MoFe proteins from *K. pneumoniae* and *R. japonicum* but not with those of *B. polymyxa* and *C. pasteurianum* (*67*). The crosses are not generally predictable and are not reciprocal in all cases. The enzyme hybrids show equal reactivity in various nitrogenase-catalyzed reactions, although some deviation from this rule has been observed by *Kelly* (*68*). Contrary to the widespread interchangeability of nitrogenase components from different organisms, no active hybrids have been found with the clostridial MoFe protein (*69*). This invites speculation that tryptophan may be involved in the formation of an active nitrogenase complex, *e.g.* by participating in the binding of the two components or in promoting the intermolecular electron transfer between them.

No amino-acid sequences are yet available, but some data from end-group analysis have been reported for the clostridial proteins. The *N*-terminus of the Fe protein contains Met-Glu(Glu, Val, Ile, Leu, Leu)Asp and the *C*-terminus contains (Asp, Gly, Ile)Tyr-Met-Leu (*49*). Since these data have been obtained from a material which differs markedly in its amino-acid composition from that of a later report, they are of a rather preliminary nature; however, leucine as *C*-terminal amino acid has been confirmed (*66*). The MoFe protein has alanine and serine (*49*) or according to a more recent report, alanine and leucine (*66*) as *C*-terminal amino acids. The presence of one terminal amino acid in the Fe protein and of two in the MoFe protein agrees with the conception of a dimeric quaternary structure of the Fe protein and a tetrameric structure of the MoFe protein with only two different subunits (see Section IV. B).

B. Molecular Weight and Quarternary Structure

The MoFe protein is a multimeric complex of high molecular weight. Data from several biological sources now seem to yield a generalized pattern. The molecular weight of the MoFe protein from *K. pneumoniae* is slightly above 200000 as determined by several independent techniques (*44*). Determination of the molecular weight of clostridial MoFe protein gave a figure of 200000 by gel filtration (*58, 70*), 221800 from the amino-acid composition (*66*), and 220000 from the subunit composition (*58*). Earlier evaluations of its molecular weight by gel filtration (*30*) or sedimentation equilibrium (*64*) have produced smaller figures which seem to be in error. The molecular weight of the rhizobial MoFe protein was found to be 182000 and 194000 by gel filtration (*33, 71*) or 197600 and 202000 by the Yphantis and La Bar method of sedimentation equilibrium, respectively (*45*). Archibald's approach to equilibrium and Yphantis' method placed the *Azotobacter* MoFe protein in the molecular weight range of 270000 to 290000 (*48*). A value of 270000 was calculated from electron micrographs (*72*) and 300000 from the molybdenum content (*39*). However, unpublished results by *Burris* (*11*) reported a smaller molecular weight of 212000 for the MoFe protein from *A. vinelandii*, which has recently been corroborated by *Kleiner* and *Chen* (*41*). Thus the overall data allow, with reasonable certainty, the conclusion that the molecular weight of the MoFe protein of nitrogenases is in the range of 200000 to 220000 daltons.

During investigation of the molecular weight of the clostridial MoFe protein, it became clear that this protein was composed of subunits. Two different types of subunits were found with molecular weights of about 50000 and 60000, determined either by polyacrylamide gel electrophoresis in the presence of dodecyl sulfate (*70*) or by gel filtration in 6 M guanidine hydrochloride (*58*). The original finding that the clostridial MoFe protein was one of the rare protein complexes of trimeric structure (*70*) has later been revised to a tetrameric structure (*58*). A quarternary structure, identical to that of *C. pasteurianum*, has been shown for the MoFe protein of *K. pneumoniae* (*44*). The tetrameric principle also extends to the MoFe protein of *Azotobacter*, with an apparently homopolymeric composition in *A. vinelandii* (*41*), but heteropolymeric in *A. chroococcum* (*62*). Since part of these conclusions was based only on dodecyl sulfate polycrylamide gel electrophoresis, more corroborating evidence is still needed. Rhizobial MoFe protein is also a tetrameric complex, with a single type of subunit of molecular weight 55000 (*45*). Serine was found as the only terminal amino acid (*33*). Thus the homopolymeric rhizobial MoFe protein might indicate a reduction in genetic material under the very special circumstances of obligatory symbiosis, requiring only two genes to code for both nitrogenase components (see below), whereas at least three genes are necessary for the more complex nitrogenase of free-living organisms. Still, following these lines, it will be interesting to see how the same catalytic mechanism is realized with four identical polypeptides or elsewhere with two pairs of different units.

The subunits of acid-treated clostridial MoFe protein have been separated by gel filtration (*58*) and their amino-acid composition has been determined (*66*). No information is available yet on the metal distribution with respect to the subunits, since the separation procedure quantitatively removes both iron and molybdenum; however, there is preliminary evidence that molybdenum is held by a dimeric protein complex (*58*).

Electron microscopic studies of the MoFe protein of *A. vinelandii* placed the four subunits in the corners of a square with dimensions of 90 × 90 Å (*73*). A tetrad of cyclic symmetry is an unusual structure, but has previously been shown for tryptophanase (*74*) and for pyruvate carboxylase (*75*). The subunit dimensions are 45 × 45 × 40 Å for which a molecular weight of 67000 was estimated (*73*).

The Fe protein is the low-molecular-weight component of nitrogenase. *Moustafa* and *Mortenson* (*50*) determined the molecular weight of the clostridial component to be 40000. Using the same technique, *Nakos* and *Mortenson* (*43*) reported a value of 55000 ± 5000. Treatment with dodecyl sulfate and mercaptoethanol dissociates the protein into two identical subunits each of molecular weight 27500 ± 1350 (*43*). Two identical subunits also compose the Fe proteins from *K. pneumoniae* (*44*). *A. vinelandii* (*41, 72*), *A. chroococcum* (*62*), and *R. lupini* (*33*). The molecular weight of the dimeric complex from these organisms with 64000 to 67000 daltons is significantly higher than that of *Clostridium.*

C. Spectroscopic Properties

Optical spectra. The two nitrogenase proteins are dark brown to yellowish in more dilute solution due to their iron-sulfur protein nature. In the dithionite-reduced state the proteins are featureless over the visible range of the spectrum (*44, 50, 64, 76*). On oxidation, the protein absorbance increases again rather indistinctly over a broad range of the spectrum. Oxidized states of these proteins are often obtained by exposure to air, a procedure that leads to inactivation and has limited physiological relevance. Recently, attempts have been made to define the oxidized spectrum of the clostridial Fe protein. Based on EPR data which revealed a physiological oxidized state of the Fe protein (see Section V. D), spectro-photometric experiments were performed in a similar manner. A complete reaction system, containing only catalytic amounts of the MoFe protein, was allowed to exhaust the re-ductant, causing the Fe protein to proceed to its oxidized state (*76, 77*). In such a way oxidized Fe protein exhibited a broad absorption maximum between 375 and 400 nm. The maximal absorbance change occurred around 430 nm with an approximate extinction coefficient of 4500 M^{-1} cm^{-1} for this transition (*76*). Some resemblance of this spectrum with bacterial ferredoxin has been noted (*76*), and may be extended to the reduced form of the high potential iron-sulfur protein of *Chromatium*. However, due to the superficial resemblance, no structural implication concerning the arrangement of the four Fe ~ S* groups of the Fe protein shall be attempted. The extinction coefficient of reduced Fe protein of *K. pneumoniae* is 4500 M^{-1} cm^{-1}, which increases to 10000 M^{-1} cm^{-1} upon air oxidation (*44*). No change in the absorbance of dithionite-reduced MoFe protein has been observed during enzyme turnover (*76*). Since further reduction of the MoFe protein beyond the dithionite-reduced state is assumed to occur during the catalytic cycle with formation of a highly reduced species (see Section V. D), apparently no absorbance is contributed by this state to the optical spectrum. Air oxidation of the MoFe protein in-creases the absorbance over a broad range with the most pronounced effect around 430 nm (*44, 64*). The extinction coefficient of the *Klebsiella* MoFe protein increases at this wavelength from 35 mM^{-1} cm^{-1} to 50 mM^{-1} cm^{-1} (*44*). Clostridial MoFe protein shows small shifts in its absorption spectrum between 300 and 400 nm at pH values below 7.2 (*78*).

In contrast to other iron-sulfur proteins, nitrogenase proteins show no circular dichro-ism in the visible range of the spectrum (*44, 66*). The Fe proteins of *Klebsiella* and *Clostri-dium* each have a pronounced trough at 220 nm, with the corresponding mean residual ellipticity indicating an α-helix content of about 33% and 13%, respectively. An α-helix content higher than that of *Clostridium* is also found in the MoFe protein of *Klebsiella*. After air oxidation – the first step towards denaturation – the α-helix content of the Fe protein, but not of the MoFe protein, decreases; a relationship to physiological events presumably does not exist. The circular dichroism spectrum of the clostridial MoFe protein may have been affected to an unspecified magnitude by the inactive species, which could have contributed up to half of the total protein according to the method of protein purifi-cation employed (*49*). This may also account for the strikingly low helix content, contrast-ing with a low tryptophan content of this protein.

EPR spectra. In contrast to the meagre information obtainable by optical spectroscopy, both nitrogenase proteins exhibit characteristic electron paramagnetic resonance spectra in the presence of dithionite. The EPR signals of the Fe protein and the MoFe protein are extremely temperature sensitive, comparable to those of other iron-sulfur proteins. Spectra are usually recorded between 10 and 20 K due to the poor resolution at higher temperatures. The dithionite-reduced Fe protein has a $g = 1.94$ type EPR signal, reminiscent of reduced plant-type ferredoxin (Fig. 3). The Fe proteins of *K. pneumoniae* (*54*) and *A. vinelandii* (*55*) have similar spectra, although slight variations in the g-values exist. Double integration of the EPR spectrum of the clostridial Fe protein relative to a Cu-EDTA standard (spectra recorded at 13 K and nonsaturating microwave power, < 10 mW) gave 0.79 electrons per four atoms of iron (*55*), or 0.2 electrons per dimeric protein molecule (*53*). The latter value, obtained by standardization at 20 K related to Cu-EDTA, was recently found to be one electron per protein molecule, or higher, depending on the protein concentration of the sample and its enzymatic activity (*56*). *Klebsiella* Fe protein yielded 0.45 electrons per protein molecule, however, the standardization temperature was not specified. This value was independent of the protein concentration and the enzymatic activity of the sample (*54*). Another small resonance was still present in the Fe protein in the $g = 4$ region, and has been attributed to some minor species of iron, either unspecifically bound or due

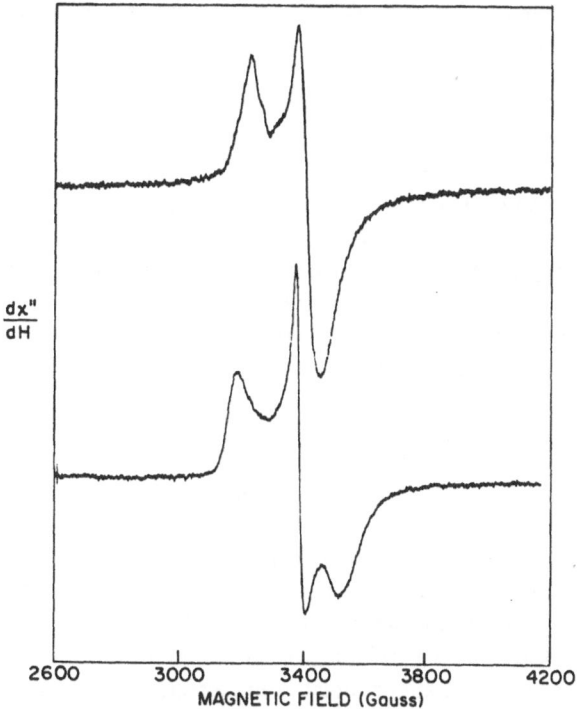

Fig. 3. Effect of MgATP^{2-} on the EPR spectrum of dithionite reduced Fe protein. Top trace, a 0.355 mM solution of the Fe protein in the presence of MgATP^{2-} (2.3 mM); g-values, 2.04 and 1.93; bottom trace, the same solution of the Fe protein without MgATP^{2-}; g-values, 2.06, 1.94, and 1.87. Spectrometer settings: sweep rate 250 gauss/min; gain 100; otherwise as indicated in Fig. 2. Reproduced by permission of Academic Press from Ref. (*116*).

15

to some protein denaturation (*53, 55*). However, since this signal becomes more prominent upon complex formation of Mg · ATP with the Fe protein, it might be necessary in further studies to heed this signal more closely. ATP binding of the protein has been related to a conformational change of the protein (*79*), and the new signal may reflect an altered environment of the iron-sulfur group(s) (*55*).

The dithionite-reduced MoFe protein has a complex EPR spectrum of a rhombic type (Fig. 2). The proteins from *A. vinelandii* (*48, 80*), *C. pasteurianum* (*53, 55*), *K. pneumoniae* (*44, 81*), and *Chromatium* strain D (*35*) all possess the same spectrum with closely related *g*-values. The MoFe protein signal is so characteristic that nitrogenase could be identified by EPR spectroscopy in whole cells of *A. vinelandii* and other bacteria (*80*), allowing the enzyme to be followed during repression and derepression. Several investigators have found an additional 1.94 type resonance in MoFe proteins of varying provenance (see Fig. 2). It has already been emphasized (Section III. C) that this signal is due to a separate species of the MoFe protein with a lower iron-sulfur content and an almost complete lack of molybdenum.

Successive oxidation by ferricyanide produces a series of complex changes in the EPR spectrum of the MoFe protein (*53*). First, the low-field signal around $g = 4$ decreases and an isotropic signal appears in the $g = 2$ region. This is followed by a rhombic and finally axial signal upon further oxidation, both of which are in the $g = 2$ region. The protein retains its catalytic activity during this oxidation, however, none of these states have been observed under physiological conditions and their significance has to await information about the structure of the iron-sulfur centers of the protein. The observed signals presumably reflect magnetic coupling of several iron-sulfur clusters at various oxidation states for which no assignment is yet possible. The MoFe protein can also be oxidized by methylene blue, phenazine methosulfate, and thionin (3,7-diaminophenothiazonium chloride) under concomitant loss of its EPR signal. With these dyes, however, the complex sequence of new EPR signals that is characteristic for the oxidation by ferricyanide, is not observed (*190*). Similarly, the dithionite-reduced Fe protein can be oxidized by ferricyanide or dyes to an EPR-silent form (*56, 190*). Further discussion of the dependence of the EPR signals upon the redox state of the nitrogenase proteins can be found in Section V. D.

The structure of the chromophores of the Fe protein and the MoFe protein, apart from the general knowledge of being iron-sulfur centers, still eludes us. The EPR spectrum of the Fe protein represents an $S = 1/2$ state. The four iron atoms might have a tetrahedrical arrangement similar to that of bacterial ferredoxin or the high potential iron-sulfur protein of *Chromatium*, or alternatively, they might represent a spin-coupled dimer of two Fe ~ S* groups similar to spinach ferredoxin (*53*). The resonance of the MoFe protein presumably arises from an $S = 3/2$ system with both rhombic and axial zero-field splittings. Since mononuclear iron is rarely found in an $S = 3/2$ state, the rhombic resonance presumably originates from a polynuclear system of spin-coupled iron clusters. This assumption is also supported by Mössbauer spectroscopy, which is reported below. Rather enigmatic is the lack of a molybdenum signal, which will be discussed in more detail in Section VI. A.

Mössbauer spectra. Mössbauer data on the *Klebsiella* nitrogenase proteins are available due to the pioneering efforts of *Kelly, Lang* and *Smith* (*61, 82*). The original investigation by *Kelly* and *Lang* (*61*) has now been improved with higher purified proteins and under better anaerobic conditions. Some of the earlier findings apparently were due to uncontrolled effects of oxidation but can be explained now in the light of recent data (82). In general, Mössbauer data support the presence of polynuclear iron clusters in the MoFe protein. The 17 to 18 iron atoms of the *Klebsiella* MoFe protein have been preliminarily distributed in three species, called M 4, M 5, and M 6. M 4 has an isomer shift of δ = 0.65 mm/s and an energy separation of ΔE = 3.05 mm/s at 4.2 K. The species is attributed to a high-spin ferrous spin-coupled pair of iron atoms. M 5 is characterized by δ = 0.61 mm/s and ΔE = 0.83 mm/s at 77 K. The area of the spectrum corresponds to eight atoms of iron of the protein molecule, which might be arranged in (Fe $\sim$ S*)$_4$ or (Fe $\sim$ S*)$_2$ clusters of ferrous iron in the low-spin state. Finally, the M 6 species is attributed to eight atoms of iron with δ = 0.37 mm/s and ΔE = 71 mm/s at 77 K (82). The spectral features of the M 6 species are similar to those of the oxidized high potential iron-sulfur protein for which a cubane structure of the four Fe $\sim$ S* groups has been shown (83). Upon oxidation with thionine, species M 4, M 5, and M 6 change into species M 1 with parameters δ = 0.37 mm/s, ΔE = 0.75 mm/s at 77 K, and M 2 characterized by δ = 0.35 mm/s, ΔE = 0.9 mm/s at 4.2 K. This oxidation is equivalent to that yielding an EPR-silent state of the MoFe protein. Both the M 1 and M 2 species are probably due to spin-coupled pairs of iron atoms or more complex clusters. The Fe protein of *K. pneumoniae* does not indicate the presence of different types of iron atoms but indicates a cluster similar to reduced clostridial ferredoxin (δ = 0.47 mm/s, ΔE = 0.95 mm/s at 195 K). At 77 K the originally symmetric doublet shows an asymmetric component at δ = 1.3 mm/s; at 4 K the spectrum considerably broadens into a multiplet, indicating the magnetic character of the system. The Mössbauer spectrum of oxidized Fe protein, equivalent to a state without EPR spectrum, could not be obtained; only air oxidized and therefore inactivated material was studied (82). Some preliminary Mössbauer data are also available from the nitrogenase proteins from *A. vinelandii* (7, 48, 72). These Mössbauer data depend very much on the purity and defined oxidation state of the material being used, as *Smith* and *Lang* (82) have convincingly demonstrated. Some of the minor Mössbauer species may still be due to protein impurities (iron proteins related or unrelated to nitrogenase) or derive from partially denatured material. A definite separation of different iron species and assignment of their structure will be a future task. It has not been possible so far to exchange externally added ^{57}Fe with protein-bound iron. A tremendous advantage would be gained through the accomplishment of iron exchange, which would, in particular, eliminate the tedious and costly way of obtaining the nitrogenase proteins for these studies from ^{57}Fe-cultured bacteria.

D. Oxygen and Cold Inactivation

Oxygen sensitivity is a common phenomenon among iron-sulfur proteins (84–86) and is also characteristic of both nitrogenase components. The Fe protein shows enhanced oxygen sensitivity compared to other iron-sulfur proteins; half lives in air of 20 s have been found for the clostridial Fe protein (unpublished) and 30 s for the Fe protein of *K. pneumoniae* (44). An inactivation of 30% was observed within 30 s with the Fe protein of *A. vinelandii*

(40). A detailed study of the oxygen sensitivity of nitrogenase is not available, but one can presume a mechanism of oxygen inactivation similar to that of the simpler iron-sulfur proteins. The inorganic sulfide of spinach ferredoxin, for instance, is oxidized to polysulfide which remains covalently bound to the protein; strong reducing agents release it as S^{2-} (87). Reconstitution of the iron-sulfur center from a sulfur zero state requires only Fe^{3+} and dithiothreitol; no additional sulfur source is necessary. The activity of the Fe protein from soya-bean bacteroids is routinely enhanced by 50% to 100% by treatment with Fe^{3+} and dithiothreitol (88). The resulting specific activity of 26 nmoles ethylene produced per min per mg protein is low compared to activities of unreconstituted Fe proteins from other biological sources, but this is the only example where partial reconstitution has been achieved with a nitrogenase protein. The particulate nitrogenase from *A. vinelandii* is inhibited noncompetitively by dioxygen (89). Its oxygen sensitivity is increased in the presence of nucleotides which may be due in part to a conformational change of the protein (90) or due to the lower oxidation-reduction potential of the nucleotide-complexed Fe protein (56). The dissociation of the Fe protein from the nitrogenase complex has also been suggested as a possible mechanism of the enhanced oxygen sensitivity in the presence of a nucleotide (91). Air exposure increases the absorbance of the Fe protein (40, 44, 50, 77) and of the MoFe protein (44, 64) between 380 and 600 nm, indicating the oxidation of iron-sulfur groups. Sulfide is lost from the air-oxidized clostridial Fe protein (50) and subsequent reduction by dithionite "bleaches" the protein beyond the original absorbance (40, 77).

The MoFe protein is more stable in air than the Fe protein and short periods of exposure do not usually affect its catalytic activity. Drastic oxidation destroys the MoFe protein and removes part of the iron as ferric oxide or oxyhydroxide (82).

The extreme oxygen sensitivity of the nitrogenase components requires that organisms which fix dinitrogen in an aerobic environment protect nitrogenase from contact with dioxygen. The *Azotobacteriaceae* are known for their vigorous respiration. Oxygen-scavenging by respiration has been proposed as one hypothesis for the protection of nitrogenase (92, 93). In support of this idea, it was shown that ammonia-grown cells can tolerate a much higher pO_2 than N_2-fixing cells. Carbon or phosphate-limited cultures were even more sensitive to dioxygen when grown on N_2, but not when grown on ammonia as a nitrogen source. It appears that aerobically N_2-fixing *Azotobacter* regulates its respiration according to its nutritional status and external O_2 tension to prevent dioxygen from reaching the N_2-fixing site. In further accordance with this proposal are observations of an increase in the number of internal vesicles of N_2-fixing cells of *A. vinelandii* (94). *Marcus* and *Kaneshiro* (95) also found a higher content of phospholipids in N_2-grown cells compared to NH_4^+-grown cells. *Swank* and *Burris* (96) noted an increase in the cytochrome c level under N_2-fixing conditions by a factor of 2 to 3, and urea as N-source compared to N_2 resulted in differences in the cytochrome a_2 content (97). However, there are also findings which weaken the evidence for a respiratory protection mechanism for nitrogenase. *Pate et al.* (98) could not confirm the results by *Oppenheim* and *Marcus* (94) and suggested instead that the different internal membrane network is due to differences in O_2 supply rather than a consequence of the N-source. Nitrogen-source dependent changes of the phospholipid content (99) and the cytochrome level (92) were also contradicted.

Nevertheless, the *Azotobacteriaceae* do fix dinitrogen in an aerobic environment and must possess a protective mechanism to ensure this. Since all experimental evidence cannot be explained by respiratory protection, a "conformational protection" has also been proposed (*92*). Because particulate nitrogenase from *Azotobacter* is rather stable in air, whereas the separated components exhibit oxygen sensitivity comparable to those of nitrogenase proteins from anaerobic organisms, structural changes of the enzyme are believed to occur that protect the oxygen-sensitive sites of the enzyme from damage. However, as a consequence of this conformational change, N_2 fixation and acetylene reduction cease as well.

Whether or not protection of nitrogenase from dioxygen is provided via respiratory control, via a "conformational" mechanism, or via a combination of both which has yet to be revealed, or other mechanisms, the problem of compartmentation of O_2-sensitive sites and O_2-requiring sites in the living cell is obvious. A rather extreme example of compartmentation might have developed in heterocystous blue-green algae. Unlike photosynthetic bacteria, blue-green algae evolve dioxygen from the water-splitting photosystem II of photosynthesis, thus jeopardizing the N_2-fixing system. Presumably nitrogenase is localized in specialized cells, the heterocysts, which do not have photosystem II, and the dinitrogen fixed there may be transported from these loci to the vegetative cells. Evidence for such a mechanism is still controversial and the reader is directed to a more complete coverage of this topic (*23*).

Cold inactivation of nitrogenase was observed in an early stage of enzymatic work (*100*) and has later been attributed to the Fe protein (*50*). After incubation for 24 hrs at 0–5 °C the clostridial Fe protein is unable to reconstitute active nitrogenase (*101*). A structural change of the protein at low temperature was indicated by a different accessibility of chelators to protein-bound iron (*50*). Cold lability has been observed with partially purified nitrogenase from several bacteria (*31, 52*) and blue-green algae (*102*), but it was recently denied for the purified Fe protein of *K. pneumoniae* (*44*). Evidence for cold lability of nitrogen-fixing activities of root-nodule bacteroids is equivocal (*32, 103, 104*) and the Fe protein from *R. japonicum* apparently can be prepared in the cold (*105*). It remains to be established whether certain Fe proteins are stable at low temperature and others are not, or whether additional factors in partially purified samples, or factors fortuitously introduced, are responsible for cold inactivation.

Cold inactivation of multimeric enzymes has been related to a weakening of hydrophobic bonding (*106*). This effect might account for the dissociation into subunits of cold labile enzymes such as carbamyl-phosphate synthetase (*107*), ATPase (*108*), pyruvate carboxylase (*109*), argininosuccinase (*110*), glucose-6-phosphate dehydrogenase (*111*), and phosphofructokinase (*112*). Cold inactivation often involves complex transitions in the conformational state of the enzyme prior to dissociation into the enzyme protomers. Sometimes the process is controlled not only by general agents, such as ionic strength and pH, but also by enzyme metabolites (*109, 112*). ATP, a negative effector of the fructokinase reaction markedly enhances cold lability, whereas enzyme substrates effectively protect against inactivation (*112*).

The Fe protein of nitrogenase occurs in solution as a dimer, so that dissociation into its subunits at low temperature appears as a reasonable hypothesis for cold lability. The

19

inactivation process of the Fe protein is irreversible, whereas other enzymes regain their activity upon rewarming. Phosphofructokinase requires a positive effector, such as fructose diphosphate, and a reductant, dithiothreitol, for reactivation (*112*). It is now well estab- lished that the Fe protein of nitrogenase binds the reaction metabolites ATP and ADP, and both might exert a still unknown effect on the cold lability of this protein. One should also attempt rewarming of the Fe protein under conditions which reconstitute iron-sulfur centers. Certainly, a reinvestigation of oxygen and cold inactivation of the Fe protein under some of the conditions suggested here is merited.

V. Enzymatic Mechanism

A. Reconstituted Nitrogenase

Preparative methods for nitrogenase yield, for the most part, the MoFe protein separated from the Fe protein and as a consequence the functional enzyme has to be reconstituted from its moieties. Any nitrogenase mechanism must take into account the composition of a functional enzyme complex and the implication resulting thereof. However, formation, composition, and stability of this complex are poorly documented. To the best of my knowledge, *Burris* (*113*) first followed dinitrogen reduction and P_i release as a function of a varying ratio of the two components. It was assumed that at maximal enzymatic activity the ratio of the two proteins will indicate the composition of the functional enzyme. From this technique *Vandecasteele* and *Burris* (*114*) reported a complex composition of 2.1 molecules of Fe protein per one molecule of MoFe protein (recalculation of these numbers on the basis of the now more accurately known molecular weight of the components will not change this result, since both molecular weights deviated by a corresponding percentage from the present data). Thus, nitrogenase would contain two molecules of the dimeric species of the Fe protein and one molecule of the tetrameric species of the MoFe protein. Recent data obtained with highly purified samples supported this ratio (*51*). A 2 : 1 com- position was also favored by preliminary data from Mortenson's laboratory (*69, 115*). *Zumft et al.* (*116*) found optimal assay conditions for the clostridial Fe protein at a molar ratio of two Fe protein per one MoFe protein. Sigmoidal activity curves resulted from titrations of a constant amount of MoFe protein with Fe protein (*69*), whereas the inverse situation resulted in inhibition when the concentration of the MoFe protein surpassed a two-fold molar excess over the Fe protein (*51, 69*). Both observations were interpreted as indicative of a non-functional nitrogenase complex with the two components balanced at equimolarity (*69, 115*). *Bergersen* and *Turner* (*88*) followed the apparent maximal velocity (V') of ammonia and ethylene formation with nitrogenase from soya-bean bacteroids as a function of increasing Fe protein. Both reactions showed sigmoidal kinetics of V' versus the Fe protein. Hill plots of these data gave interaction coefficients of about 2, suggesting that the MoFe protein has interdependent binding sites for the Fe protein.

The findings of *Eady* (*117*) and *Eady et al.* (*44*) are in contrast to this. Titrations of the Fe protein from *K. pneumoniae* with its MoFe protein showed a rapid activity increase for H_2 evolution, acetylene and dinitrogen reduction, with a sharp maximum at an equimolar ratio of the two proteins. Further increase of the MoFe protein concentration inhibited all three activities, similar to findings with the clostridial system. Differences in the shape of titration curves were observed with cyanide (*28, 118*) and methyl isonitrile reduction (*68*), both requiring for optimal activity a higher amount of the MoFe protein than C_2H_2 or N_2 reduction. Equimolar mixtures of the nitrogenase of *K. pneumoniae* also gave a single boundary in the ultracentrifuge in the absence of dithionite. A sedimentation coefficient of 12.4 S was found for the boundary of the enzyme complex in comparison with 10.5 S of free MoFe protein under the same conditions. Increasing the protein ratio beyond 1 : 1 produced an additional boundary with a sedimentation coefficient of 4.5 S, equalling that of free Fe protein (*117*).

A sharp activity optimum of titration curves at integral protein ratios implies rather tight binding of the two components. This was indicated by ultracentrifugation studies with an essentially symmetrical boundary of the sedimenting complex (*117*). On the other hand, although the overall protein ratio is known in titration experiments, no information of the actually bound protein is given, and will depend especially upon the catalytic activity of the nitrogenase components. To illustrate this with *Klebsiella* nitrogenase, with a presumed equimolar composition, it was shown by *Thorneley* (*47*), who monitored an absorbance change of the operating enzyme at 425 nm, that maximal activity occurred at a MoFe protein to Fe protein ratio of 1 : 3 or higher. Although this ratio would indicate less active material than one with maximal activity at an equimolar ratio, it was of significantly higher specific activity (the MoFe protein was almost twice as high) than that giving apparently the minimal active composition of the nitrogenase complex. Until complementing techniques yield more information on the composition of an active nitrogenase complex, titration data of steady-state enzymatic activity must be viewed with caution. It has been obvious for some time that a large excess of the Fe protein is required for maximal activity of the MoFe protein (*44, 51, 116, 119*). Contrary to this well-established fact, there appears to be a tight complex of the two nitrogenase components, for which a dissociation constant, $K_d < 0.5 \mu M$ has been claimed (*47*). Hence, *Ljones* (*76*) had already suggested a very fast formation and dissociation of the nitrogenase complex, faster perhaps than the electron transport itself. In addition, *Zumft et al.* (*56*) concluded from rapid-freeze studies of the decrease of the low-field EPR signal of the MoFe protein, that the interaction among the proteins may be faster than the electron transfer. The reduction of the MoFe protein was thus visualized to occur by combination with free reduced Fe protein in solution and immediate dissociation thereafter (*76*).

A further problem is the assignment of catalytic activity to the tetrameric or dimeric species of the MoFe protein. In solution, the MoFe protein exists preferentially as a tetramer. Partial dissociation into smaller molecular weight units has been observed with the clostridial MoFe protein (*58*) only at low protein concentrations (< 5 mg/ml). Such low protein concentrations are used under assay conditions, but the influence of the Fe protein and the other reactants on the quarternary structure of the MoFe protein − if any, what kind, and to what extent − are not known. Since the MoFe protein contains two minimal units, each with one atom of molybdenum, each one of these minimal units could already be catalytically active. Alternatively, although a tetrameric quarternary structure of the MoFe protein might be a prerequisite for catalytic activity, the protein might still have more than a single active center. No experiments have yet been designed to prove or disprove these possibilities.

B. Substrates and Kinetic Studies

Seven classes of substrates are reduced by nitrogenase (Table 4). In all cases both protein components are required, together with reductant and ATP. The basic compounds of these substrates are N_2, N_2O, N_3^-, C_2H_2, HCN, CH_3CN, and H_3O^+. With the exception of the electron transfer to hydronium ions, all substrate reductions are inhibited by CO.

Molecular nitrogen is the physiological substrate of nitrogenase. Dinitrogen takes up six electrons and is reduced to two molecules of ammonia (*103, 120, 121*). No end-product inhibition has been observed *in vitro*, although ammonia, in a mechanism which has yet to be revealed, represses the enzyme of N_2-fixing cells (*57, 122−126*, see Section X.). The affinity of nitrogenase for N_2 is higher than for all other substrates. The K_m for N_2 ranges from 0.03 to 0.1 mM (*17, 36, 37, 103, 104, 127−131*). A striking difference exists between the K_{N_2} of intact cells of *A. vinelandii* and that of the isolated enzyme. Whereas the former is only about 0.01 to 0.02 atm (*130, 132*), values of 0.1 to 0.2 atm have been found *in vitro* (*36, 37*). This has been considered as a physiological advantage due to a presumably low pN_2 inside the cell (*36*), but *Bergersen* and *Turner* (*88*) showed that the apparent Michaelis constant of rhizobial nitrogenase depended on the relative ratio of the two

Table 4. Nitrogenase-catalyzed reactions

Substrates	Products	Ref.	K_m [a])	Ref.
N_2	NH_3	(*24, 118*)	0.10 ± 0.03 atm	(*150*)
N_2O	N_2, H_2O, NH_3 (?)	(*133, 137*)	0.05 atm	(*129*)
N_3^-	N_2, NH_3	(*129, 136, 339*)	$(1.4 \pm 0.1) \times 10^{-3}$ M	(*150*)
HCN	CH_4, NH_3, C_2H_4, C_2H_6, CH_3NH_2 (?)	(*129, 146*)	1.3×10^{-3} M	(*36*)
CH_3NC	CH_4, CH_3NH_2, C_2H_6, C_2H_4, C_3H_8, C_3H_6	(*146, 147*)	2×10^{-3} M	(*36*)
H_3O^+	H_2	(*27, 120, 151*)	−	−
ATP	ADP, P_i	(*120, 157, 158*)	$1-3 \times 10^{-4}$ M	(*36, 162*)
$S_2O_4^{2-}$	SO_3^{2-}	(*119, 340*)	$8-9 \times 10^{-3}$ M	(*36, 160*)

[a]) These values vary considerably and other figures can be found in the literature.

nitrogenase proteins and suggested that a failure to match the *in vivo* concentrations in artificial systems could account for the differences in K_m. Although *Azotobacter* nitrogenase is obtained as a complex, selective loss of one component might have occurred during the preparation procedure. I would like to emphasize this example of an enzyme with a marked difference in behavior inside the bacterial cell and isolated from it. Without regard to the possible explanation of mismatching the *in vivo* conditions, it may hint more importantly at regulatory properties of the nitrogenase system which have no counterpart *in vitro*.

Nitrogenase catalyzes no isotope exchange, $^{14}N_2 + ^{15}N_2 \rightleftharpoons 2\,^{14}N^{15}N$, indicating no initial cleavage of dinitrogen at the active site of the enzyme (*11, 24*). Attempts to detect free intermediates have been unsuccessful, but diimine and hydrazine have repeatedly been proposed as enzyme-bound intermediary states of N_2 reduction (*129, 133, 134*). Recently *Bulen* (*135*) found that hydrazine, preferentially in its unprotonated form, is a substrate of *A. vinelandii* nitrogenase, lending strong support to its supposed role as an intermediate.

Azide is reduced to equal amounts of NH_3 and N_2 by a two electron transfer (*129, 136*). Of the N_2 formed, only about 7% is further reduced to NH_3. At low azide concentrations and a low gas-to-liquid ratio in the assay system, to yield a higher pN_2, about 29% of the N_2 is reduced further. This indicates that the N_2 formed from N_3^- is not in the correct binding position to undergo immediate further reduction (*129*). $K_{N_3^-}$ values of 1 to 1.3×10^{-3} M have been reported (*36, 129, 136*), but a value of 2×10^{-4} M (*136*) has also been reported. Nitrous oxide is reduced by two electrons to N_2 and H_2O (*137*). Similar to azide reduction, the N_2 formed from N_2O is not reduced to NH_3 without subsequent interaction with the enzyme (*129*). The K_{N_2O} is 0.05 atm (*129*). The reduction of NO by nitrogenase is not conclusive (*129, 138*); nitric oxide inhibits N_2 fixation competitively at a very low level ($K_i = 4.3 \times 10^{-7}$ M) but the observed N_2 might originate from N_2O which has been formed non-enzymatically by dithionite from NO (*129*).

The triple bond of acetylene is reduced partially by a two electron transfer to give ethylene (*139*). Other reductions of this type include $CH_3C\equiv CH \rightarrow C_3H_6$ and $C_2H_5C\equiv CH \rightarrow C_4H_8$, but $CH_3C\equiv CCH_3$ and $CH_2=CHCH=CH_2$ are not reduced (*140*). The reduction of acetylene has become the routine assay for nitrogenase either in the laboratory of for field studies (*13, 141*). The reaction is stereospecific with *cis*-1,2-deutero-ethylene being found in the presence of D_2O (*13, 68, 139*). A K_m for C_2H_2 has been reported in the range of 0.01 to 0.03 atm for *A. vinelandii, C. pasteurianum*, and soya-bean root nodules (*36, 139, 141*). Concentrations above 0.2 atm are inhibitory; at 1 atm compared to the rate at 0.2 atm, ATP hydrolysis is diminished by 44% and C_2H_2 reduction by 24% (*36*).

ATP-dependent dihydrogen evolution from H_3O^+ (*134*) is the last in the series of two-electron transfer reactions catalyzed by nitrogenase. In the absence of any externally added reducible substrate, all electrons are used for H_2 production. With other substrates present, nitrogenase still evolves dihydrogen to a varying degree which is only suppressed in the presence of C_2H_2. Acetylene thus competes best for electrons (*48*). At saturating pN_2, approximately 25% of the electrons go to H_2 and 75% to NH_3 formation (*38*), but with any electron partition, the transfer rate remains constant (*13, 27, 129*).

A peculiarity of nitrogenase is the formation of HD, either from D_2 and H_2O or from H_2 and D_2O. This reaction was originally found in intact soya-bean root nodules by *Hoch et al.* (*133*) and was later confirmed and extended to cell-free preparations (*142, 143*). HD formation is enhanced in the presence of dinitrogen, implying that HD is produced by exchange of D_2 with hydrogen of enzyme-bound intermediates like diimine or hydrazine (*133*). HD formation in extracts of *A. vinelandii* was also stimulated by N_2 and had the same requirements as for N_2 fixation, *i. e.* ATP and reductant (*134*). Carbon monoxide was inhibitory and no HD was formed in an atmosphere of argon or with acetylene or cyanide as substrates. HD formation by nitrogenase was assumed to be similar to the exchange reaction catalyzed by aryldiimine- and arylhydrazine-Pt complexes (*134*). Most of these results were confirmed by *Kelly* (*144*) with nodules from *Medicago lupulina*, but an increased HD formation was not observed in N_2 with cell-free extracts of *A. vinelandii. Burris* (*11*) then indicated that the participation of hydrogenase could not be excluded in these experiments and *Vandecasteele* and *Burris* (*114*) could find little HD formation from D_2 with the purified clostridial nitrogenase proteins either in N_2 or Ar. Despite these discrepancies, most of the experimental evidence still supports an increased HD formation in N_2. *Bulen* (*135*) has recently advanced a new mechanism. Electron-balance studies led him to the conclusion that D_2 competes for the hydrogen atoms of an enzyme-bound intermediate (presumably diimine). However, rather than being a true exchange reaction, HD is formed by decomposition of the intermediate, yielding N_2 again (N_2 being thus catalytic in the cycle) and consuming a surplus of electrons. This mechanism would also explain the inhibition of N_2 fixation by dihydrogen (*130, 138*). Consistent with this model is the observation that ATP-dependent H_2 evolution is not inhibited by H_2 (*145*), arguing that inhibition of N_2 fixation involves a different site than H_2 evolution.

An ample variety of compounds with CN triple bonds is reduced by nitrogenase. The most simple compound, cyanide, is reduced by six electrons to about equal amounts of CH_4, NH_3, and about 10% CH_3NH_2 (*129*). In one instance, in addition to CH_4, traces of ethylene and ethane in the proportion $CH_4 : C_2H_4 : C_2H_6 = 100 : 0.08 : 0.07$ have been reported (*146*). Methyl isonitrile is reduced to methane, ethylene, and ethane in the proportion of $100 : 0.28 : 2.1$ (*146*). Methylamine and ethylamine are additional reduction products but contrary to *Hardy et al.* (*147*) no C_3 products were observed (*28*). Table 5 gives a survey of various nitriles, isonitriles, and alkynes that are reduced by nitrogenase. Not all reactions have been studied to the same depth, leaving some discrepancies with respect to product identification and product ratio. Only a few of the listed compounds can be regarded as enzyme substrates. Most of them have a large K_m with extremely low reduction rates, and only their hydrocarbon products are detectable by gas chromatography.

From the variety of substrates utilizable by nitrogenase, some conclusions can be drawn with respect to binding, orientation, and stereochemical effects. Evidence from chemical models allows the assumption that substrate binding involves metal-binding sites. All substrates have triple or potential triple bonds which permit π-bonding (side-on orientation) and they have one or more pairs of non-bonding electrons which allow σ-bonding (end-on orientation) (*129*). Only C_2H_2 has no free electron pairs and side-on bonding is indicated by the stereospecific *cis*-addition of hydrogen (*13, 68, 139*). Both types of bonding can lead to a bridged complex between two binding sites (*148*) or to a single end-on (side-on) binding at a mononuclear site. Again, mononuclear end-on bonding to yield an acetylide-type bonding is not probable for C_2H_2 as indicated by the *cis*-product

Table 5. Reduction of nitriles, isonitriles, and alkynes by nitrogenase

Substrates	Products	Electrons	K_m	Relative rates [a]
Nitriles				
HCN	CH_4, C_2H_6, C_2H_4, NH_3	4, 6	0.4–1.0	0.5
CH_3CN	C_2H_6	6	500–1000	0.004
C_2H_5CN	C_3H_8	6	500–1000	0.003
CH_2CHCN	C_3H_6, C_3H_8, NH_3	6, 8	10–25	0.2
cis-$CH_3CHCHCN$	1-butene, cis-2-butene, n-butane, trans-2-butene	6, 8	100–200	0.007
trans-$CH_3CHCHCN$	trans-2-butene, 1-butene, n-butane, (NH_3)	6, 8		0.0007
CH_2CHCH_2CN	1-butene	6, 8		0.003
$CH_2C(CH_3)CN$	isobutylene	6		0.0003
C_4H_9CN	n-butane	6		0.0002
$(CN_3)_2CHCN$	not reduced			
Isonitriles				
CH_3NC	CH_4, C_2H_6, C_2H_4, C_3H_6, C_3H_8, CH_3NH_2	6, 8, 10, 12, 14	0.2–1.0	0.8
C_2H_5NC	CH_4, C_2H_6, C_2H_4, $C_2H_5NH_2$	6, 8, 10	10–25	0.2
CH_2CHNC	CH_4, C_2H_4, C_2H_6			
Alkynes				
C_2H_2	C_2H_4	2	0.1–0.3	4
CH_3CCH	C_3H_6	2		
CH_2CCH_2	C_3H_6			
CH_3CCCH_3	not reduced			
C_2H_4	not reduced			
$CH_2CHCHCH_2$	nor reduced			

[a]) Rates compared with dinitrogen reduction. Adapted from Ref. (147).

formed in D_2O. Alkynes are not reduced beyond their corresponding alkenes and presumably due to steric effects, dimethylacetylene is not reduced. Equally inert are substrates with olefinic double bonds, either isolated or in conjugation. Allene, in contrast, is reduced but probably after isomerization to methylacetylene (147).

Isonitriles are, in general, better substrates than nitriles, indicated by a smaller K_m and an increased reduction rate of the former (131, 147). End-on bonding at the carbon atom is likely, considering that methyl isonitrile is reduced at a substantially higher rate than acetonitrile, that methylamine is a reaction product of methyl isonitrile reduction, and that CD_4 is formed in D_2O (146, 147). The formation of C_2 or C_3 hydrocarbons of methyl isonitrile has been related to interaction of vicinal C_1 radicals at a dinuclear binding site (146) or to an alternating insertion-reduction mechanism (147). In the nitrile series, reduction is observed up to n-butyronitrile, but the branched molecule, isobutyronitrile, cannot be reduced (140). Introduction of olefinic bonds into nitriles and isonitriles considerably increased the affinity of a compound to nitrogenase (131, 140). Interaction of these molecules at a second site on nitrogenase has been proposed. An intensive study has been performed with acrylonitrile, the reduction of which shows several interesting features (149). The compound is either reduced by six electrons to propylene and ammonia, or by eight electrons to propane and ammonia. The formation of propylene involves the migration of a double bond as indicated by the following equation:

$$CH_2{=}CHC{\equiv}N + 6\,e^- + 6\,D^+ \rightarrow CH_2DCH{=}CD_2 + ND_3\,.$$

Likewise, a double-bond shift occurs during the reduction of cis- and trans-crotonitrile (149). Acrylonitrile and crotonitrile yield saturated hydrocarbons as reaction products, indicating that the only olefinic groups reduced by nitrogenase have to be in conjugation with a $C{\equiv}N$ bond. In D_2O, nitrile reduction is increased by a factor of two to five, whereas other substrates show no isotopic effect (149). There is some evidence that nitrogenase has more than one site for binding and reduction of this variety of substrates. Inhibition studies by Hwang et al. (150) were explained by assuming a minimum of five sites. These are: (i) H_3O^+ site, (ii) CO site, (iii) N_2 site, (iv) C_2H_2 site, and (v) azide, cyanide, and methyl isonitrile site. Carbon monoxide divides the enzyme generally into two parts (129, 151). One represents the site of ATP-dependent electron activation where H_2 is evolved and which is unaffected by CO, the other represents the substrate-reducing part of the enzyme, the activity of which is completely blocked by CO. This division of nitrogenase, however, does not coincide with the Fe protein and the MoFe protein. The mechanism of CO inhibition is still equivocal; competitive as well as noncompetitive inhibition of N_2 fixation has been observed (138, 150). The presence of multiple sites was already suggested by the stoichiometry of product formation from azide and nitrous oxide. Although nitrogenase has a much higher affinity for N_2 than for N_3^- or N_2O, azide reduction yields about equimolar quantities of N_2 and NH_3, whereas N_2O reduction yields about 30 times as much N_2 as NH_3 (129, 136, 137). This high ratio in favor of N_2 production is indicative that the N_2 site is different from that of N_3^- or N_2O. The inhibition of N_2 reduction by dihydrogen, but not that of azide or nitrous oxide supports this view further (150). The inhibition mode of H_2 has been described as competitive and also noncompetitive. These data possibly require reevaluation in the light of the inhibition mechanism mentioned above in connection with HD formation. However, they leave little doubt that the N_2 binding site is a separate locus on nitrogenase. Inhibition of N_2 reduction by C_2H_2 is noncompetitive (151), revising a previous contrasting report by the same research group. Different ATP requirements are observed for N_2 and C_2H_2 reduction (36). The C_2H_2 site, therefore, seems to be distinct from the N_2 site. Cyanide and methyl isonitrile are competitive inhibitors of azide reduction. Cyanide in turn is noncompetitive with N_2 reduction, and acetylene and azide are noncompetitive with each other. This places HCN, N_3^-, and CH_3NC in one group, which is separate from the N_2 site and the C_2H_2 site (150). Different sites for each C_2H_2, HCN, and N_2 reduction were also indicated from kinetic data by *Eady et al.* (*118*). Figure 4 depicts graphically the proposed multiple-site structure of nitrogenase as outlined above.

Most of the experimental evidence suggests that these sites are located on the MoFe protein. Replacement of Mo by V (see Section VI. A) resulting in altered substrate affinity (*152, 153*), and indirect evidence from chemical models (*6*) indicate an involvement of Mo and therefore the MoFe protein in substrate binding. The electron-transfer sequence derived from EPR and Mössbauer spectroscopy indicates that the MoFe protein acts as a terminal electron acceptor from where reducing equivalents are transferred to substrates. The appearance of a new EPR signal of the MoFe protein in the presence of CO (*154*) indicates that CO interacts with this protein. Since CO inhibits substrate reductions, it would follow that the substrate binding sites are on the MoFe protein. However, the CO-induced EPR signals in the MoFe protein still need to be confirmed by other laboratories. The apparent K_m values for N_2 and C_2H_2 reduction are affected by variation of the Fe protein concentration but not by that of the MoFe protein (*88*). This again places the substrate binding sites on the MoFe protein. An elegant experiment, yielding strong direct support for placing the substrate reducing site for C_2H_2 on the MoFe protein has been performed by *Smith et al.* (*54*). The EPR signals of the dithionite-reduced MoFe protein from *Klebsiella* showed shifts in line position over the pH range 6.8 to 9.5 (g-values of the

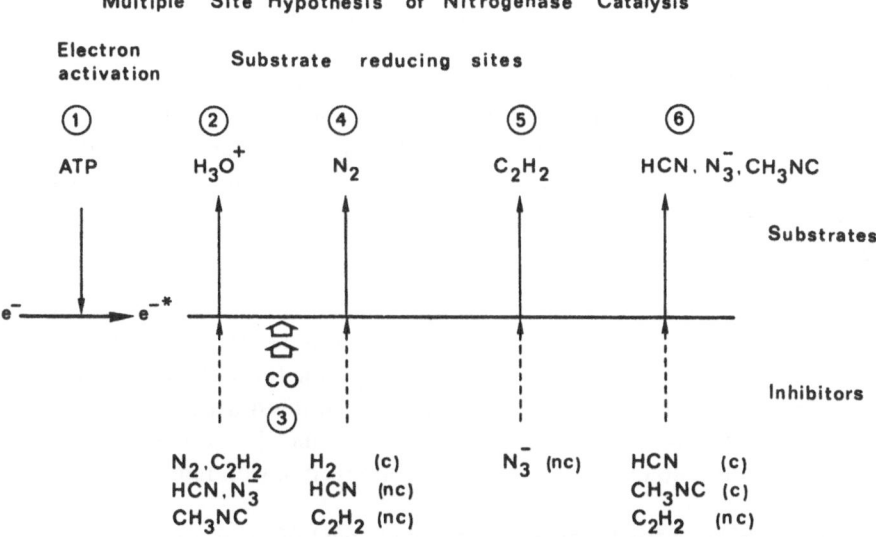

Fig. 4. The multiple-site hypothesis of nitrogenase-catalyzed reductions. Carbon monoxide divides the enzyme into an electron activating part and a substrate reducing part. Carbon monoxide also clearly separates the H_2 evolution site from the other substrate reducing sites. The mutual inhibition among the substrates, with the type of inhibition indicated in the lower half (c, competitive inhibition; nc, noncompetitive inhibition), separates the N_2 site from the C_2H_2 site and these in turn from the HCN, N_3^-, and CH_3NC site. The data are taken from Refs. (*113*), (*150*), and (*151*).

low-pH form were g_1 = 4.32, g_2 = 3.63, and g_3 = 2.009, those of the high-pH form were g_1 = 4.27, g_2 = 3.78, and g_3 = 2.018). The effect had a pK of 8.7 and was most pronounced at the g_2 feature. In the presence of acetylene at pH 8.2, displacement of the low-pH form in favor of the high-pH form occurred, indicating binding of C_2H_2 to the MoFe protein and competition of this compound for protons. EPR experiments by *Evans et al.* (*155*), which suggested an interaction of cyanide with the MoFe protein, could not be confirmed by other laboratories. However, binding studies by co-chromatography involved the MoFe protein but not the Fe protein in cyanide binding (*156*).

C. The Role of ATP

One of the most intriguing but perhaps least resolved problems of the nitrogenase reaction is the mechanism of ATP action. The involvement of ATP in N_2 fixation has been inferred from the inhibition by arsenate and glucokinase, the latter competing with the ATP-generating system for phosphate. The stepwise elaboration of the concept that the ATP requirement of N_2-fixing cell-free extracts is due to direct nucleotide-enzyme interaction, has been treated in a previous review (*11*). *Hardy* and *Knight* (*157*) indicated that ATP hydrolysis was accompanied by P_i release and *Kennedy et al.* (*158*) established that this was stoichiometric to ADP formation with no other products being formed. Nitrogenase also has an obligatory requirement for a divalent cation. *Burns* and *Bulen* (*145*) noticed the need for Mg^{2+} with ATP-dependent dihydrogen evolution of *A. vinelandii* extracts. *Dilworth et al.* (*17*) then found that a replacement of Mg^{2+} by Mn^{2+} did not affect the ATP-generating system but decreased N_2 fixation by 70% and attributed its effect to the nitrogenase system. The same authors found that Ca^{2+} was inhibitory and that Co^{2+} and Fe^{2+} affected the energy supply system more strongly than the N_2-fixing system. *Hardy* and *Knight* (*157*) extended these observations to reductant-dependent ATP hydrolysis. The need of the N_2-fixing system for a divalent metal acting elsewhere than on the energy generating system was further corroborated with the separated nitrogenase proteins and substrate amounts of ATP (*159*). The optimal Mg^{2+}/ATP ratio with substrate levels of ATP was shown to be 0.5 (*158–161*). *Burns* (*160*) systematically investigated the cation specificity of *A. vinelandii* nitrogenase with dihydrogen evolution as an assay system. As indicated by earlier results, nitrogenase is not absolutely specific for Mg^{2+}, but can use Mn^{2+} almost equally effectively, and Co^{2+}, Fe^{2+}, and Ni^{2+} in this order of decreasing efficiency. The replacement of magnesium by a paramagnetic metal is of potential interest for further studies to elucidate the mechanism of ATP interaction with resonance techniques.

Kennedy et al. (*158*) suggested that the metal requirement of the nitrogenase reaction was due to the formation of a reactive metal-nucleotide complex. Binding of Mg·ATP to the clostridial nitrogenase components was shown by co-chromatography on Sephadex G-50 to be restricted to the Fe protein (*156*). ADP was also bound to the Fe protein, but unlike with ATP, no Mg^{2+} ions were required. Approximately 0.4 mole of nucleotide was bound to the protein (*156*). In an effort to repeat these experiments, it was shown by co-chromatography that both nitrogenase proteins of *K. pneumoniae* bound Mg·ATP; likewise, the nucleotide was bound by air-inactivated Fe protein, cytochrome c, and clostridial ferredoxin. Equilibrium dialysis proved to be unsuccessful in demonstrating any interaction and hence the nucleotide binding effect was assumed to be unspecific (*162*).

However, support for the binding capacity of the Fe protein came from work with *A. chroococcum* which confirmed the earlier result with *C. pasteurianum*, but unfortunately was not extended to the MoFe protein (*90*). Binding experiments were recently refined by the use of a gel equilibration technique other than co-chromatography (*163*) and by EPR spectroscopy (*79, 116*). The rhombic EPR signal of the Fe protein changes in the presence of ATP and Mg^{2+} into a nearly axial-type signal (Fig. 3) with *g*-values of 2.04 and 1.93 (*116*). This signal change is accompanied by a line broadening and decrease of the amplitude (major resonance) of approximately 22%, whereas the spin concentration remains constant (*79*). Other authors also observed a decrease of the spin population of 22% (*55*), which might be due to general difficulties in quantitating the Fe protein signal (*53*). *Zumft et al.* (*79, 116*), proposed an ATP-induced conformational change of the Fe protein, to explain the transition of the EPR signal in analogy to a similar effect in the presence of 5 M urea. Titration of this transition with Mg·ATP indicated the binding of two ATP molecules to the dimeric Fe protein. Two identical non-cooperative binding sites for ATP on the Fe protein, with a dissociation constant K_d of 17 μM, were found by the gel equilibration method (*163*) and a requirement of Mg^{2+} for the binding of either ATP or ADP, as detected earlier by EPR spectroscopy, was confirmed. However, for Mg·ADP only one strong binding site was found with a dissociation constant, $K_d = 5$ μM, which is considerably smaller than that for ATP (*163*). The proposal of an ATP-induced conformational change of the Fe protein was further supported by following the oxidation-reduction potential of this iron-sulfur protein in the absence and in the presence of the nucleotide [(*56*), see Section V.D)]. A conformational change was also indicated by an enhanced —SH group reactivity (*164*) and an increased accessibility of protein-bound iron to 2,2'-dipyridyl (*165*) in the presence of Mg·ATP. Nucleotides other than Mg·ATP are inactive in the latter effect, and Mg·ADP even protects the Fe protein from reaction with the chelator (*166*). This contrasts with the observation by EPR spectroscopy, showing Mg·ADP is operative at high concentrations in the spectral transition. The action of nucleotides in the presence of dipyridyl probably cannot be interpreted in terms of the free nucleotide. Nucleotides form complexes with 2,2'-dipyridyl (*167*), which may render the system quite complicated through the presence of various equilibria, and the active species ought to be identified in each case.

The clear requirement of magnesium ions for all nucleotide effects, including requirement for catalysis, is convincingly explained by complex formation of the divalent metal with the nucleotide to establish the configuration necessary for nucleotide protein interaction. The structure of ATP in solution has recently caused renewed interest through investigation by NMR (*168, 169*). The most likely structure appears to be an outer-sphere-type complex with a water molecule bridging the metal and the N-7 atom of the adenine ring (*168*). The requirement of a distinct nucleotide structure for interaction with nitrogenase is demonstrated by substitutions at the purine moiety or the introduction of a methylene group in $\alpha-\beta$ position of the phosphate chain, which render ATP inactive to induce the change of the EPR spectrum (*79*). At cation concentrations in excess of the nucleotide, the formation of a dimetal-ATP complex has been indicated [see (170) and references therein]. This complex and the free nucleotide are inactive or even inhibitory in the nitrogenase reaction (171, 172). No attention has previously been paid to the equilibria among Mg^{2+}, ATP^{4-}, $Mg \cdot ATP^{2-}$, and $Mg_2 \cdot ATP$, although it was known for a while that a high Mg^{2+} concentration is inhibitory for nitrogenase (158). Recently, this situation has changed and some of the older data have been criticized because of this negligence (171, 172). Still, to augment the complexity of nitrogenase, magnesium ions do not only intervene in the equilibria of nucleotide complex-formation, but they also seem to have direct effects on the binding of the two enzyme components (171).

No binding of $Mg \cdot ATP$ to the MoFe protein was observed with equilibration techniques (156, 163), nor was the EPR spectrum affected (79). *Orme-Johnson et al.* (55), however, found a decline of 26% in the signal height of the MoFe protein with added $Mg \cdot ATP$. Since it was shown that extremely small quantities of the Fe protein still allow a turnover of the system (56), a contribution of the Fe protein to the observed signal decline appears as a likely explanation.

ADP is a potent inhibitor of the nitrogenase reaction with a $K_i = 3.5$ mM (120, 159, 161). AMP, in contrast, inhibits only acetylene reduction of cell-free extracts by 50% to 60%, but not of purified systems (159, 161). Hill plot coefficients for the interaction of ATP with the operating nitrogenase close to two have been reported for *C. pasteurianum* (159), and *A. vinelandii* (38), however, a value of only one has been reported for *K. pneumoniae* (161). In the presence of ADP the coefficient for the clostridial system increased to about three (159). This suggests a more complex mechanism of nucleotide interaction at the binding sites of the enzyme than at the isolated component which shows no cooperativity for $Mg \cdot ATP$ binding. However, it is of interest in this context that $Mg \cdot ADP$ slightly enhances $Mg \cdot ATP$ binding to the Fe protein (163). Deviation from Michaelis-Menten kinetics of the total electron flow in the presence of ADP has been noticed (76), and it has been proposed that the modulation of nitrogenase activity *in vivo* is regulated via ADP (159, 161). To overcome any inhibition of nitrogenase by ADP *in vitro*, phosphatidokinase ATP-generating systems are used routinely (128, 173).

ATP requirement and concomitant hydrolysis are observed with all reactions of nitrogenase: dinitrogen reduction (128, 173, 174), dihydrogen evolution (120, 145, 151, 175), reduction of acetylene (139), and reduction of nitrous oxide (137), azide (136), cyanide (129), isonitrile (146), and acrylonitrile (149). Other nucleoside triphosphates, *e.g.* GTP, UTP, or CTP, cannot replace ATP (13, 159). The apparent K_m for ATP-dependent acetylene reduction with *Klebsiella* nitrogenase was 1.2×10^{-4} M (162). The K_m of ATP-supported H_2 evolution of *A. vinelandii* was 3×10^{-4} M (160) and similar values have been obtained for N_2 reduction, P_i release, and total electron flow (36). A K_m of 1.8×10^{-4} has been found for clostridial nitrogenase (76).

Inhibitors of electron transport and oxidative phosphorylation such as gramicidin, rotenone, antimycin A, oligomycin, CCCP, dinitrophenol, and ouabain did not inhibit ATP hydrolysis by nitrogenase in concentrations which completely block other energy transducing processes (157, 158).

The term "reductant-dependent ATPase" has been selected for the ATP hydrolyzing part of nitrogenase to distinguish this activity from the classical ATPase, and to emphasize that ATP utilization is coupled to electron transport. This coupling of ATP breakdown to electron transport has resulted in considerable efforts to establish the number of ATP molecules required per minimum of two electrons transferred. Since a single orthophosphate is eliminated from ATP, this relationship shall be designated as the P/e_2 ratio. The following basic properties of this ratio have been established: (i) no single value can be ascribed to this ratio, (ii) the ratio is temperature (37) and pH (78, 176) dependent, and (iii) ATP is hydrolyzed by nitrogenase to some extent in the absence of reductant (78, 177). Most workers have found a P/e_2 ratio of four to five, but extremes of two (78, 157) and more than twenty (119) have also been reported. Originally a ratio of four was calculated from partially fractionated clostridial extracts (158, 178). At the same time a P/e_2 ratio of two was reported for a particulate nitrogenase of A. vinelandii (157), and also, independently from this work, a ratio of five (27) has been reported. To resolve these discrepancies Winter and Burris (176) investigated the P/e_2 ratio with ethanol-treated nitrogenase of C. pasteurianum which showed little ATP-independent H_2 evolution or reductant-independent ATPase activity. With corrections for these two activities a P/e_2 ratio of four was found at pH 6.0, which increased to 4.5 at pH 8.0. Bui and Mortenson (177) then found that nitrogenase of C. pasteurianum hydrolyzed ATP in the absence of an electron source, an effect which was particularly high below pH 6.0. Jeng et al. (78) confirmed this finding and showed that P/e_2 ratios for ATP-dependent H_2 evolution, which varied from eight to three over the pH range 5.6 to 7.6, were in reality two over the entire pH range once a correction for the reductant-independent ATP hydrolysis had been applied. The contribution of reductant-independent ATP hydrolysis to total ATP hydrolysis at pH 6.0 was around 80% in the original report (177) and around 50% in the subsequent one (78). At pH 7.0 this process shared between 50% and 30% of the overall ATP breakdown. Smaller contributions of reductant-independent ATP hydrolysis were observed by other investigators, and Hadfield and Bulen (37) attributed only 15% P_i release to reductant-independent ATP hydrolysis in A. vinelandii nitrogenase.

The data of different research groups were clearly incompatible with respect to P/e_2 values as well as the existence and degree of reductant-independent ATP consumption. Ljones and Burris (119), therefore, reinvestigated these questions with thoroughly purified clostridial nitrogenase components and found maximally 10% P_i release due to reductant-independent ATP hydrolysis (the presence of a contaminating ATPase could be ruled out with certainty). Moreover, they showed that the P/e_2 ratio was dependent on the relative ratio of the Fe protein and the MoFe protein. When the amount of Fe protein was kept constant and the MoFe protein concentration was increased, consumption of reductant showed a maximum and was inhibited on further increase of the MoFe protein. ATP hydrolysis, however, remained constant yielding P/e_2 ratios of up to twenty or higher. In the reverse experiment, when the Fe protein was increased and the MoFe protein remained constant, no inhibition of dithionite oxidation was observed and the P/e_2 ratio reached a constant number of four after initial values slightly above five (119). With A. vinelandii, apart from one early report indicating a low value (157), the work by Hadfield and Bulen (37) and Silverstein and Bulen (38) firmly supports a P/e_2 ratio of five at 30 °C. A. vinelandii nitrogenase is inherently different from, and has definite advantages over that of C. pasteurianum because the enzyme does not need to be reconstituted from its compo-

31

nents. Although *Kelly* (*28*), using nitrogenase from *A. chroococcum*, claimed that addition of acetylene or methyl isonitrile to the enzyme in an atmosphere of argon changed the P/e_2 ratio as compared to H_2 evolution under argon alone, data of *Hadfield* and *Bulen* (*37*) did not support this finding. The P/e_2 remained constant when Ar was gradually replaced by N_2, or N_2 by H_2; albeit electron allocation to protons or N_2 varied. Most striking was a dependency of electron allocation upon the ATP concentration (*38*). Though the P/e_2 ratio again remained constant, dihydrogen was preferentially evolved at low levels of ATP, while more electrons were allocated to dinitrogen only at higher ATP concentrations, and ammonia formation exceeded H_2 evolution. *Silverstein* and *Bulen* (*38*) suggested that an ATP-dependent step in the enzyme might occur repeatedly at a constant rate requiring successively higher activated enzyme molecules for the transfer of two, four, or six electrons to their respective substrates.

A branch point in the deactivation mode of a reactive enzyme-ATP complex has been postulated (*38*) from the temperature dependency of the P/e_2 ratio which increased from 4.3 at 20 °C to 5.8 at 40 °C (*37*). The assumption was that a different activation energy is necessary for a process leading to only ATP hydrolysis to that of the process which also leads to dihydrogen evolution. Such a dual mechanism might also explain the inhibition of electron transfer at a high MoFe protein concentration (without affecting ATP hydrolysis) and possibly involves enzyme complexes of different composition (*119*). Conflicting with this view, however, is the identical activation energy for N_2 fixation, ATP hydrolysis and H_2 evolution (*160*), although the data are not completely convincing for ATP hydrolysis (*37*). Below 21 °C the activation energy for these processes is 39 Kcal mole^{-1}, above this temperature it is 14.6 Kcal mole^{-1} (*160*). If a dual mechanism of ATP hydrolysis exists, a search not only for a single P/e_2 ratio but also of integer numbers for the overall reaction will be in vain.

Some of the remaining incoherencies with respect to the P/e_2 ratio will be resolved as further progress is made. Yet, the question arises whether dinitrogen fixation is inherently an energy-wasting process. In intact N_2-fixing cells of *C. pasteurianum*, 20 moles of ATP are consumed per mole N_2 fixed (*179*); for *K. pneumoniae* this value is 30 (*180*), but only five for *A. chroococcum* (*181*). At present one can only speculate whether a tight coupling of ATP hydrolysis to electron transfer is possible to give a P/e_2 ratio of one or two; the experimental evidence is against it.

Several possibilities of ATP action have been proposed. Activation of nitrogenase has been envisaged to occur by phosphorylation of the enzyme (*128*) or by conversion into a form of highly negative oxidation-reduction potential (*78, 151, 157, 182*). The latter is often called "electron activation" and the resulting potential has been liberally placed at − 400 mV (*157*) or at − 700 mV (*183*). ATP-induced conformational changes at the substrate-reducing site have been proposed to bring metal coordination centers into an appropriate position for substrate binding (*148*), or to change the ligand configuration of a transition metal, such as to create the reductive species (*173*). The action of ATP as proton source at a non-aqueous dinitrogen reducing site has been suggested (*176*), as well as the intervention of solvated electrons (*177*). The electron-activation hypothesis has gained impressive support by the recent demonstration of a low potential form of the Fe protein in the presence of ATP (*56*) and investigation of the electron-transfer sequence, where the formation of a highly reduced species of the MoFe protein has been indicated (*54, 55*). The maximal redox potential change of the Fe protein of 110 mV will contribute only modestly to the total energy requirement and further activation must be involved, presumably through the combined action of ATP and the Fe protein at a site on the MoFe protein. At present it is not known which step in the catalytic cycle produces ATP hydrolysis that might be related to the formation of a highly reduced form of nitrogenase. One asks how the chemical energy of ATP is transferred to the protein and is utilized for substrate reduction? Most suggestions are too vague to serve as a working hypothesis and more definite models for ATP action are required. In this context, it is of interest that molybdothiolborohydride complexes show favorable catalytic activity in the presence of ATP (*184, 185*), albeit the mechanism of ATP is also obscure in these cases (but see Section VII).

D. Oxidation-Reduction Properties

Oxidation-reduction potentials. The redox potentials of the Fe protein and the MoFe protein have been investigated by a technique combining potentiometry with low temperature EPR spectroscopy (*56*). Because of the distinct EPR spectra of these two proteins, which are susceptible to oxidants and reductants, this technique appears to be the method of choice rather than a spectrophotometric method with which attempts have been made to determine the redox potential of the two proteins (*77*). The former technique has definite advantages. Oxidized and reduced states of the nitrogenase proteins are readily distinguishable by EPR, since the oxidized state is completely void of a spectrum and the reduced state can be obtained with dithionite. Full reduction is achieved when addition of reductant does not further increase the EPR signal. The changes in the EPR spectrum during the redox transition are simple; no other paramagnetic species appear in partially oxidized or reduced specimens. Quantitation of redox changes is thus facilitated by peakheight measurements. Equilibration of the proteins with the electrode is accomplished by a variety of mediating compounds. Other than in spectrophotometric experiments, spectral overlap does not present a problem as most mediators do not have an EPR spectrum or their signals can easily be distinguished from the protein spectrum. It is not necessary to know the exact redox potential of the mediators under experimental conditions because the redox potential of the entire system is measured continuously over the complete redox cycle.

By this technique, clostridial MoFe protein was shown to have a redox potential of approximately − 20 mV at pH 7.5 (*56*). A one-electron transfer was involved in the redox step (Fig. 5). With the same technique, a redox potential around − 60 mV at pH 7.5 was found for the MoFe protein of *Chromatium* (*186*). This protein also showed a second, low potential center around − 260 mV which was destroyed by ferricyanide with concomitant loss of most of the catalytic activity of the protein. Clostridial MoFe protein did not exhibit such a low potential component. Since dithionite reduces the MoFe protein of *Clostridium* and *Klebsiella* only sluggishly, incomplete equilibration as a result may cause the biphasic nature of the redox titration. Despite this possibility, the MoFe protein may

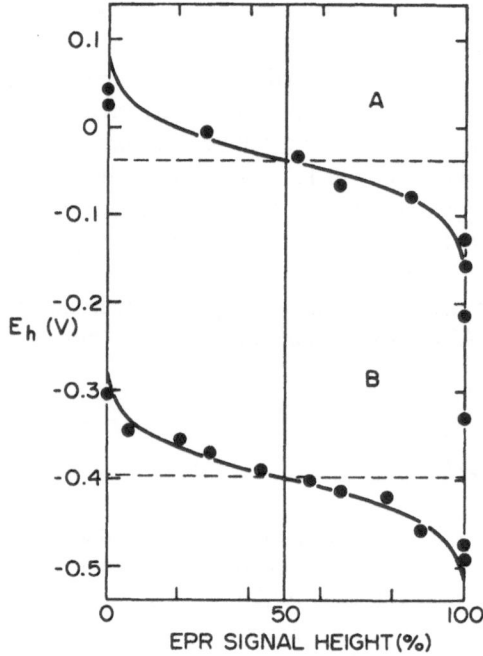

Fig. 5. Redox titration of the MoFe protein and its demolybdo form. (A) Titration of the MoFe protein (70 μM) in a solution of nine mediators (E_{m7} range -325 to $+135$ mV) with a saturated solution of thionine. The $g = 3.78$ resonance was plotted. (B) Titration of the demolybdo form of the MoFe protein (150 μM, mol. wt 110 000) in a solution of seven mediators (E_{m7} range -444 to -137 mV) with ferricyanide. The $g = 1.92$ resonance was plotted. The titration was reversible to 65%. For A and B the buffer was 0.1 M TES-KOH, pH 7.5. The MoFe protein has a midpoint potential of -33 mV, $n = 0.97$; the demolybdo form has a midpoint potential of -395 mV, $n = 0.99$. The solid lines are the corresponding theoretical one-electron transitions. Reproduced by permission of the Fed. of Eur. Biochem. Soc. from Ref. (56).

indeed have a low potential component as indicated by the titration of the inactive species of the MoFe protein. This species, which has preserved some of the structural features of the active molecule (42), has a redox potential of approximately -395 mV at pH 7.5 but no high potential component (Fig. 5). Absence of this negative component in titrations of the MoFe protein indicates that it must be masked by more prominent groups, or that its properties are altered to a high potential center upon incorporation of molybdenum and the missing part of iron.

A redox titration of the clostridial Fe protein giving a potential around -300 mV is shown in Fig. 6. Again only one electron was involved in the redox step. The potential was independent of the pH within ± 20 mV. A peculiarity of the clostridial Fe protein is that it is only stable in a redox environment lower than -200 mV. Whenever the potential is raised above this threshold, immediate and irreversible inactivation occurs (56). The electrochemical properties of the nitrogenase proteins are summarized in Table 6.

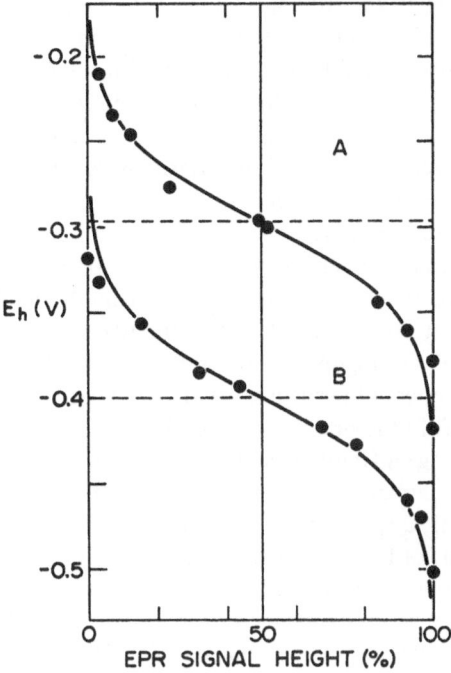

Fig. 6. ATP-induced shift of the redox potential of the Fe protein. (A) The Fe protein (180 μM) in a solution of seven mediators (E_{m7} range -444 to -137 mV) was titrated oxidatively with ferricyanide. The $g = 1.87$ resonance was plotted. The titration was reversible to 80%. The midpoint potential is -297 mV, $n = 1.01$. (B) The same system titrated in the presence of 2 mM ATP and 2 mM MgCl$_2$. The $g = 1.93$ resonance was plotted. The titration was reversible to 90%. The midpoint potential is -400 mV, $n = 1.04$. For A and B the buffer was 0.1 M TES-KOH, pH 7.5. The solid lines are theoretical one-electron transitions calculated on the basis of the indicated midpoint potentials. Reproduced by permission of the Fed. of Eur. Biochem. Soc. from Ref. (56).

Table 6. Electrochemical properties of the nitrogenase proteins

	Fe protein	MoFe protein	Demolybdo MoFe protein
Dithionite-reduced state	paramagnetic	paramagnetic	paramagnetic
Oxidized state	EPR-silent	EPR-silent	EPR-silent
Electrons involved in the redox transition	1	1	1
Redox potential $E_{m7.5}$	-300 mV	$-$ 20 mV -260 mV	-395 mV

Data are taken from Refs. (56) and (186).

35

The transition induced by Mg·ATP in the EPR spectrum of the Fe protein, which was related to a conformational change, could also be related to a change in the redox potential (56). In the presence of Mg·ATP the clostridial Fe protein titrated reversibly within a range of -371 to -437 mV (pH 7.5; Fig. 6). Since the Fe protein under turnover conditions of the nitrogenase reaction is present in its ATP form, this lower redox potential appears to be effective in the catalytic cycle of the enzyme. Alternatively, the redox potential shift could be caused by preferential binding of Mg·ATP to the oxidized Fe protein or by a similar interaction with a mediator. Either alternative, however, is unlikely. The value of 110 mV for the redox potential shift may still change upon closer examination of the intervening factors and refinement of the employed methods. Especially disatisfying in the original method was the lack of a direct protein transfer from the titration vessel to the EPR tube. Since one deals with a species of highly negative redox potential, any O_2 introduced during the transfer procedure would result in a positively deviating redox potential, and hence a smaller potential shift in the presence of Mg·ATP. Since its discovery by EPR spectroscopy in early 1973, the Mg·ATP-induced change of the redox potential of the Fe protein has subsequently been demonstrated with similar results, by adopting the original potentiometric method to optical spectrometry (187).

Knowledge of the redox potentials of both nitrogenase proteins allows one to conclude formally an electron transfer sequence in the order: primary reductant → Fe protein → MoFe protein. Such a sequence was also deduced from kinetic EPR studies of the complete nitrogenase system with the main difference that the high-potential component of the MoFe protein (around -20 mV) was not thought to be involved rather than a further, highly reduced species of this protein (see below). This species is only formed in the presence of the Fe protein, Mg·ATP, and reductant, and hence its redox potential is not amenable to measurement by classical potentiometry. A promising new technique is being developed by *Watt* (188), using electrochemically generated methyl viologen as electron donor to a complete nitrogenase system. A threshold voltage, above which dihydrogen evolution ceased, was found at -380 to -410 mV (data given versus the hydrogen electrode). Half saturation of dihydrogen evolution was at approximately -450 mV. This latter value would place nitrogenase among the most negative biological redox catalysts known. A similar negative threshold potential for acetylene reduction has been observed with *Chromatium* nitrogenase (189).

Kinetic studies. The EPR spectrum of the combined nitrogenase components is additive, except for the slightly less pronounced high-field doublet of the Fe protein (55, 190). *Orme-Johnson et al.* (55) also found a 16% decrease of the MoFe protein signal and an increase and sharpening of the Fe protein signal. These observations can be considered to be of minor significance for the following, although analysis of such changes might give much needed information on the interaction of both proteins and constitution of an active nitrogenase complex. When ATP as its Mg-complex was added to a mixture of the two nitrogenase components, a substantial decrease in the MoFe protein signal occurred (54, 55, 81, 116, 190). The decrease, which equally comprises all resonances of the MoFe protein, reflects a turnover of the system with hydronium ions as substrate. A ratio of two moles of Fe protein per one mole of MoFe protein, in the presence of an ATP generating system and an excess of dithionite (Fig. 7), caused the almost complete disappearance of the low-field resonance and the g = 2.01 spike of the clostridial MoFe protein (*55, 190*). Precisely which factors control the magnitude of this signal decrease has not yet been fully elaborated. Dependence of *Klebsiella* nitrogenase on the dithionite or Fe protein concentration has been indicated (*54, 81*). Thus, increasing the concentration of these reactants caused a corresponding signal decrease of the steady-state spectrum of nitrogenase. However, a signal decrease until about 30% to 40% of the signal remains is not comparable to the clostridial system, where under all conditions a steady-state value smaller than 15% of the remaining MoFe protein signal has been observed. For example, varying the MoFe protein to Fe protein ratio from 1 : 14 to 1 : 0.06 did not produce any qualitative differences, but

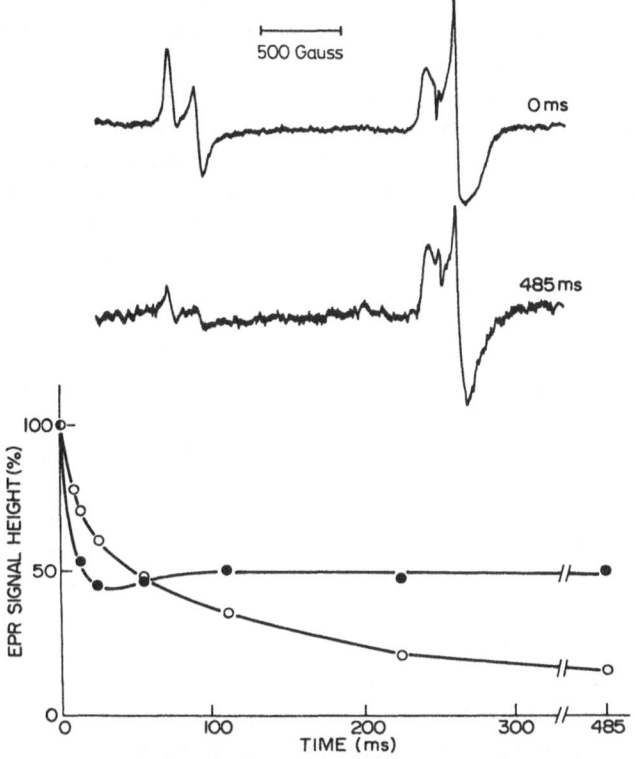

Fig. 7. Time course of the EPR signal changes of nitrogenase in the pre-steady-state. A mixture of the Fe protein (0.2 mM) and the MoFe protein (0.1 mM) was reacted for the indicated time intervals with an ATP-generating system. The reaction was carried out at room temperature, with hydronium ions as substrate. The total signal amplitude of the g = 4.27 and the g = 3.78 resonance of the MoFe protein (o) and the g = 1.94 resonance of the Fe protein (•) were plotted against time. The two top traces show the EPR spectra of reconstituted nitrogenase at 0 ms and after 485 ms of reaction time. Note the almost complete absence of the MoFe protein signal after 485 ms in contrast to the appreciable signal amplitude of the Fe protein. The spectra were recorded at gain 100 (0 ms) and gain 200 (485 ms); the temperature was 20 K and the modulation amplitude 10.5 Gauss. Reproduced by permission of the Fed. of Eur. Biochem. Soc. from Ref. (*56*).

the rate of the signal decline was slowed to a half-time of 19 s with extremely low concentrations of the Fe protein (*56, 190*).

A rapid-freeze technique has to be used to monitor the EPR signals of a nitrogenase, containing the Fe protein in excess over the MoFe protein, during the pre-steady-state. This method involves the usual rapid-mixing devices and stops the reaction by freezing the reactants instantaneously in $-140\,°C$ isopentane. The clostridial and *Klebsiella* nitrogenases have been subjected to these measurements (*54, 56, 190*). The EPR signals of the clostridial MoFe protein declined with a half-time of 50 to 70 ms (*56, 190*) and reached their steady-state level of approximately 15% remaining signal after 500 ms (Fig. 7). The mixing sequence of the three reactants, MoFe protein, Fe protein, and ATP-generating system, did not affect the time course of the MoFe protein signal. The EPR signal of the Fe protein initially showed a very fast decrease ($t_{1/2} \approx 10$ ms), then a small trough, and finally a high (50% and more of the signal remaining) and constant level, before the MoFe protein signal had reached the steady-state.

The results from an investigation on the mixing sequence of the *Klebsiella* nitrogenase were contrary to these findings (*54*). In a stopped-flow study with the same enzyme, the EPR data could not be substantiated by monitoring an absorbance change at 425 nm (*47*). Complex formation between the two nitrogenase components appears to be very fast (rate constant $k > 10^7$ M^{-1} s^{-1}) and is not detectable by the rapid-freeze method. A further difference between the clostridial and the *Klebsiella* system is the small signal decrease of the latter of only 30% to 50%, and a kinetic pattern of the Fe protein, indistinguishable from that of the MoFe protein (*54*). Claims on electron transfer rates or rate-limiting reactions should be viewed with caution as long as these discrepancies remain unresolved. However, there is an agreement on the overall phenomena of signal decline and reappearance under defined reducing or oxidizing conditions (*54, 55, 56, 190*).

When reductant is limiting and dithionite is being exhausted by dinitrogen evolution, the EPR signal of the MoFe protein returns to its original state, whereas the signal of the Fe protein disappears completely (Fig. 8). These signal changes of the nitrogenase system are interpreted as changes of the oxidation state of the proteins. An oxidative (*116, 190*) and a reductive mechanism (*54, 55, 81, 190*) have been proposed to explain the observed changes. The following arguments support a process by which the MoFe protein proceeds in the reaction to a higly reduced state:

(i) reducing equivalents in the form of the ATP-complex of the Fe protein have to be added to the reaction system to cause the decrease in the MoFe protein signal,
(ii) the MoFe signal is decreased during, or completely absent from the steady-state, and as the most forcible argument
(iii) the MoFe protein signal is recovered upon physiological oxidation of the system (*54, 55, 81*).

Following this argument, the MoFe protein as isolated and in the presence of dithionite appears to be in an "oxidized" state. It is preferable to call this state "semi-reduced" or "semi-oxidized" to distinguish it from a more oxidized state which is reached by dye-oxidation (*81, 116, 190*). The MoFe protein has an EPR signal only in its semi-reduced form, which is lost either on reduction during the catalytic cycle or upon dye-oxidation. The

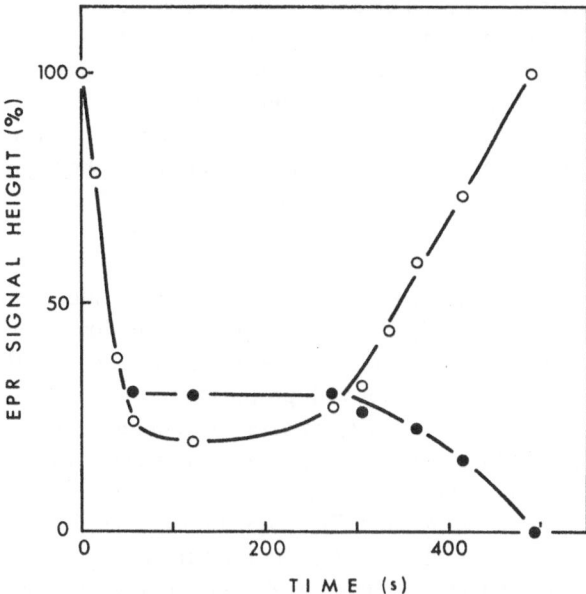

Fig. 8. EPR signals of a reconstituted nitrogenase during turnover under reductant-limiting conditions. The reaction mixture was composed of 0.1 mM MoFe protein, 0.005 mM Fe protein, 5 mM ATP, 10 mM $MgCl_2$, 10 mM creatine phosphate, 100 units/ml creatine kinase, 100 mM Tris-HCl (pH 7.5), and dithionite as present in the protein solutions (approx. 1 mM). The EPR spectrum of a 0.3 ml sample was recorded at the indicated time intervals. Changes in the g = 3.78 signal of the MoFe protein ($\circ$) and the g = 1.94 signal of the Fe protein ($\bullet$) are shown. The signal of the Fe protein was corrected for some spectral contribution of the demolybdo form of the MoFe protein. After 300 s when dithionite becomes exhausted, the signal of the Fe protein disappears, whereas that of the MoFe protein returns to its original intensity. The zero-time spectrum was obtained prior to addition of ATP. The low steady state intensity of the Fe protein signal might be a result of the difficulty in quantitating a small amount of this protein in the presence of excess of the MoFe protein (*W. G. Zumft, L. E. Mortenson* and *G. Palmer,* unpublished).

accumulation of the reduced form of the MoFe protein without significant enzyme turnover at extremely low levels of the Fe protein has been related to a direct involvement of the Fe protein-ATP complex in substrate reduction (*56*). Thus, the mere formation of the reduced, EPR-silent form of the MoFe protein does not appear sufficient on its own for substrate reduction.

Initially, when the MoFe protein signal was observed under reducing conditions (*i.e.* in the presence of dithionite) and an EPR signal-free state could be produced reversibly by dye-oxidation without affecting the catalytic activity of the protein, an oxidative mechanism was proposed (*116, 190*). With respect to the possible existence of a highly reduced form of the MoFe protein, it is worthwhile to analyze the situation of simpler iron-sulfur proteins. The iron-sulfur clusters of *Peptococcus aerogenes* ferredoxin are geometrically identical to the Fe_4S_4 center of the high potential iron-sulfur protein of *Chromatium* (*191*). To resolve the considerable difference of about 750 mV in the redox potential of the two structures, the existence of a total of three oxidation states rather than two for each one of these proteins has been proposed (191). Thus, bacterial ferredoxin should be able to undergo further oxidation and the high potential iron-sulfur protein further reduction. This would require that one Fe_4S_4 cluster accommodates a total of two electrons rather than one (192). The postulated positive redox potential of bacterial ferredoxin has not yet been detected, however, *Cammack* (193) found a potential of the high potential iron-sulfur protein lower than − 600 mV in 80% dimethyl sulfoxide. Since Fe_4S_4 clusters are assumed to occur in the MoFe protein, these finding may serve as a model for the occurrence of "super-reduction" in nitrogenase.

Mössbauer studies. Further evidence for the conversion of semi-reduced MoFe protein to a highly reduced form during the catalytic cycle is available from Mössbauer studies (82). Similar to the EPR experiments, a complete nitrogenase system containing ^{56}Fe protein and Mo ^{57}Fe protein, was allowed to turn over with protons as substrate, by adding reductant and an ATP-generating system. A new species, M 7, was observed in the steady state with an isomer shift δ = 0.42 mm/s and an energy separation ΔE = 1.04 mm/s at 77 K (82). Two species, M 4 and M 5, of the dithionite-reduced MoFe protein remained essentially unchanged but the species, M 6, disappeared. The magnetic species M 6 was assumed to originate from two 4 Fe ~ S* clusters which tentatively form the EPR active center of the MoFe protein (82). Its Mössbauer behavior might parallel the observations made by EPR spectroscopy, with the absence of a paramagnetic state of the MoFe protein during enzyme turnover (116). Assignment of the M 6 → M 7 transition as a reduction stems primarily from analogy with the transformation of oxidized to reduced high potential iron-sulfur protein. The parameters at 77 K for the oxidized form (δ = 0.32 mm/s; ΔE = 0.79 mm/s) and the reduced form (δ = 0.42 mm/s; ΔE = 1.12 mm/s) of this protein compare reasonably with those of the M 6 (δ = 0.37 mm/s; ΔE = 0.70 mm/s) → M 7 (δ = 0.42 mm/s; ΔE = 1.04 mm/s) transition of the MoFe protein. The isomer shift from δ = 0.37 mm/s to δ = 0.42 mm/s is consistent with a reduction, *i.e.* an increase in $3d$ electron density (82). The species M 1 and M 2, characteristic of dye-oxidized material, were not detected during the steady state. Thus, if the catalytic cycle includes the oxidized state of the MoFe protein (also assuming that physiologically and dye-oxidized proteins are identical), its concentration must be small. Different atmospheres, Ar versus N_2, or the use of acetylene did not affect the Mössbauer spectrum of nitrogenase. The conversion of species M 6 to M 7 was also unaffected in the presence of CO. This is constitent with the observation that CO does not inhibit dihydrogen evolution of nitrogenase. On the other hand, CO inhibits N_2 and C_2H_2 reduction and the formation of the catalytic active species should be prevented. It thus appears that the events observed by Mössbauer spectroscopy during enzyme turnover, like those observed by EPR, do not fully explain the catalytic cycle, nor do they identify the final active species. Let us assume that electrons are ultimately channeled to substrate via molybdenum, which undergoes a change in oxidation state, we have not seen anything of this event either by EPR or by Mössbauer spectroscopy.

Table 7 shows the three possible oxidation states of the MoFe protein. Only one redox potential is known at present, namely the redox pair oxidized−semi-reduced MoFe protein. Preliminary evidence indicates that the second redox couple, namely semi-reduced−fully reduced MoFe protein, has a potential in the vicinity of − 500 mV (188, 189).

Table 7. Proposed oxidation states of the MoFe protein

Redoxpotential	EPR signal	Oxidation state	Conditions
	no	reduced	Complete nitrogenese reaction system
< -400 mV		$\uparrow\downarrow$	
	yes	semi-reduced	Dithionite
-20 mV		$\uparrow\downarrow$	
	no	oxidized	Mild oxidants (Dyes)
0 mV		$\uparrow\downarrow$	
	yes	over-oxidized	Strong oxidants (Ferricyanide, O_2)

VI. Metals in Dinitrogen Fixation

A. Molybdenum, Vanadium, and Tungsten

It was recognized at an early stage in nitrogenase investigation that N_2-fixing microorganisms have an enhanced requirement for molybdenum which was absent from cells supplied with reduced nitrogen, such as urea or ammonia (2). Chemical analysis of purified nitrogenase has established that molybdenum is part of the MoFe protein. Molybdenum exists in oxidation states from two to six and forms a variety of complexes, including complexes with dinitrogen (8). This suggests that molybdenum plays a role in the nitrogenase reaction as a redox component and/or as a coordination center for substrates. However, the biological process has produced little evidence to support either possibility. The oxidation states of molybdenum (V) and (III) are paramagnetic species amenable to EPR spectroscopy. This latter technique has led to profound insight into the catalytic mechanism of xanthine oxidase (63, 194), but has been of little help thus far in elucidating the role of molybdenum in nitrogenase. A paramagnetic state of molybdenum might not be present in the MoFe protein as isolated, but may become evident under oxidation-reduction transitions. Again, this is not the case with nitrogenase. Any paramagnetic state must either be extremely short-lived, having escaped detection in rapid-freeze experiments or, alternatively, must be obscured by interaction with other paramagnetic centers of the protein. Even this latter possibility is limited as isotope enrichment studies have shown. Incorporation of 76% [95]Mo into the clostridial MoFe protein did not show hyperfine structures in the EPR spectrum (53). Similar negative results were observed with the MoFe proteins of *K. pneumoniae* (54) and *Chromatium* (35). Against this background, it is intriguing that *Orme-Johnson*

(*154*) claimed for nitrogenase, in the presence of CO, EPR signals around $g = 1.98$ with hyperfine splittings characteristic of molybdenum.

Molybdenum has been implicated in biological catalysis by its interchangeability with vanadium (*152, 153*) and tungsten (*91*). Evidence for the replacement of molybdenum by vanadium is indirect by altered enzyme properties, rather than by co-purification or chemical analysis of the purified protein. *Bortels*, who discovered the Mo requirement of N_2-fixing organisms, also found that V could partially satisfy the Mo requirement of *Azotobacter* (*195*). This has again been recently confirmed with *A. vinelandii* by *McKenna et al.* (*152*), who demonstrated that bacterial growth in the presence of V was better than in the absence of either Mo or V. However, no such effect exists in the blue-green algae *A. cylindrica*; V rather tends to enhance the symptoms of Mo deficiency and inhibits growth and nitrogenase activity even in the presence of normal amounts of Mo (*196*). Preliminary evidence for the involvement of Mo in nitrogenase catalysis comes from the change in the K_m for acetylene, which was 4×10^{-3} M in extracts of Mo-grown cells and 2×10^{-2} M in identical V-containing preparations (*152, 153*). Purification of the V analogue of the MoFe protein from V-grown cells resulted in a preparation with an optical spectrum, solubility, antigenicity against a MoFe protein antiserum, and amino-acid composition similar to the MoFe protein (*72*). However, V was lost during the last steps of purification and only 40% of the usual Fe content was detectable in the V analogue. The V analogue showed a high electron allocation to H_3O^+ (*153*) and a different product ratio (propylene/propane) of acrylonitrile reduction, which was 2.5 for the V analogue compared to about 6 for the MoFe protein (*149, 153*). The K_i for CO inhibition of N_2 and C_2H_2 reduction was approximately five-fold higher for the V analogue than for the MoFe protein (*153*). These results would involve molybdenum in the entire reduction process including substrate binding, electron transfer, and product dissociation (*72*). In a contradictory report, no evidence for the synthesis of a V analogue of the MoFe protein was found (*197*). *Benemann et al.* (*198*) concluded from painstaking comparison of the Mo content of the V analogue and its enzymatic activity that residual Mo could in fact account for all of the catalytic activity. The clue to this conclusion was the observation that cells cannot use Mo efficiently at extremely low Mo levels in the growth medium. This inhibition is overcome by V or a very small addition of Mo. Although nitrogenase from V-grown cells was more labile than the enzyme from a normal growth medium, the enzyme was found to be stabilized when compared to the enzyme of cells which were grown under severe Mo starvation. Thus, a structural role for V in the MoFe protein rather than a catalytic one has been proposed (*198*).

The major difficulties of metal substitutions are to prove that (i) the analogous protein is synthesized at all, (ii) the metal is indeed incorporated into the protein at the site of molybdenum, and (iii) that an observed catalytic activity is due to the incorporated metal and not due to residual Mo. Protein-specific detection techniques, *e.g.* immunological reaction with antisera should help to detect, quantitate, and follow these metal-analogue proteins during purification and analysis.

Whereas the question of V incorporation into the MoFe protein and activity of such a protein analogue remains somewhat ambiguous, experimental evidence is far better with respect to the role of tungsten. This metal has also been used to replace Mo in *Azotobacter* MoFe protein (*91*). Co-purification of W with nitrogenase activity was found to a limiting extent and the resulting enzyme preparation showed strong inhibition of C_2H_2 reduction, H_2 evolution, and ATP hydrolysis. Traces of Mo might again account for this residual activity, since the acetylene-reducing activity of the W analogue could not be confirmed (197). Molybdenum appears generally to be replaceable by W in Mo-containing enzymes such as nitrate reductase (199), sulfite oxidase (200), or xanthine oxidase (201). In all cases, the resulting enzyme analogue was catalytically inactive in its essential reaction. An intriguing and apparently singular case is the behavior of nitrogenase of R. rubrum against tungsten (202). The nitrogenase-dependent H_2 evolution of Mo-deficient cells was stimulated by a 4 h treatment with tungstate to the same extent as with molybdate. Dinitrogen fixation showed a similar, though somewhat reduced capacity to be restored by tungsten. Although it was assumed that W substitution for Mo leads to an active nitrogenase in R. rubrum, it was also stated that growth of this organism in the presence of 10^{-5} M tungstate did not exceed growth without Mo. A more detailed study will be required to clarify this point. As one possibility, it might be suggested that tungsten behaves in R. rubrum similar to V in Azobacter and in fact intervenes in Mo utilization or enzyme induction rather than exhibiting catalytic activity itself. Molybdenum is needed for enzyme synthesis in Azotobacter and Klebsiella (197, 203), a requirement which can, at least in Azotobacter, be partially satisfied by tungsten. The study of a purified W analogue of the MoFe protein could provide tremendous insight into the catalytic properties of nitrogenase. In contrast to the various effects of V and W on the MoFe protein, Azotobacter cells will synthesize a full and active complement of the Fe protein, irrespective of the type of metal supplied during growth (91, 197).

The existence of a component or cofactor common to molybdenum-containing enzymes has been proposed (204) as a result of intercistronic complementation of the impaired assimilatory nitrate reductase of the Neurospora crassa mutant nit-1 (205, 206). The original interpretation that molybdenum-containing enzymes share a common protein subunit was later narrowed (204) because nitrate reductase activity could be reconstituted by a variety of sources including the acid-treated MoFe proteins, xanthine oxidase, aldehyde oxidase, and others. The unlikelihood of an immutable protein during evolution of such different enzymes, led to the suggestion that the component may be actually a low-molecular-weight cofactor. The existence of such a cofactor was postulated earlier in genetic work on nitrate reductase from Aspergillus nidulans (207). Since the in vitro reconstitution was accompanied by an increase in the sedimentation coefficient of the complemented nitrate reductase, it was suggested that the cofactor played a role in stabilizing or producing the quarternary structure of the enzyme, in addition to possessing a presumed electron transferring property (204). A component which was dialysable and could reconstitute the nitrate reductase of the N. crassa mutant nit-1 was found in cell-free extracts of R. rubrum, denitrifying bacteria, and several species of Bacillus (208, 209). Despite its apparent low molecular weight, it could be precipitated by ammonium sulfate, had a sedimentation coefficient between 4.5 and 5.2 S, and thus might be bound in certain organisms to a larger protein moiety (209). It has now been shown that this component does in fact contain molybdenum and the conditions of the in vitro assembly of nitrate reductase have been thouroughly studied (210). Tungsten can also be incorporated into this component, yielding an inactive enzyme (211). Low molecular weight compounds of Mo (mol. wt $\approx$ 1300) have been isolated from the clostridial MoFe protein, which could possibly represent this Mo cofactor (*212*). A nitrogenase mutant UW 45 of *Azotobacter* is apparently impaired at the same site (*i.e.* the cofactor) as *Neurospora* mutant *nit*-1. Acid-treated MoFe protein or acid-treated extracts from various N_2-fixing organisms, which were previously shown to be sources of this component, could reconstitute the N_2-fixing ability of this mutant *in vitro* (*213*).

Ganelin et al. (*214*) dissociated a Mo-containing component from the MoFe protein of *Azotobacter* by exposure to 0.5 M NaCl. The component had a simple EPR signal around $g = 2$ which was attributed to Mo(V). The amino acids Arg, Asp, Thr, Ser, Glu, Gly, Ala, Val, Ile, and Leu were shown to be present. Assessment of this report depends on the purity of the source and the biological activity of the Mo component. Both criteria, however, were not clarified. If further investigations verify the existence of a low-molecular-weight cofactor of molybdenum, the consequences for any mechanism of molybdenum-containing enzymes will doubtlessly be far reaching.

B. Iron

Various spectroscopic techniques have provided ample evidence for the participation of iron in nitrogenase catalysis and most of these data have already been presented in the previous sections. A general involvement of iron is evident from the nutritional requirement of N_2-fixing organisms (*2*) and the inhibition of enzymatic activity by *o*-phenanthroline and 2,2′-dipyridyl (*157*). MoFe proteins with a low iron content, such as the demolybdo MoFe protein of *C. pasteurianum* (*42*) and the vanadium analogue of *A. vinelandii* (*72*), are catalytically inactive or show only a low activity. Evidence suggesting a stepwise removal of iron from the clostridial MoFe protein (*215*) cannot be considered as conclusive since the presence of the Mo-free inactive MoFe protein was not recognized. The iron of the reduced clostridial Fe protein appears to be quite stable against complexing with 2,2′-dipyridyl. However, in the presence of Mg · ATP, 80% of the iron was removed from a sample containing 3.3 Fe per protein dimer within 1 h, compared to only 11% during the same period in the absence of the nucleotide (*166*). A more specific role of iron in both nitrogenase components was assigned by EPR and Mössbauer spectroscopy (*54, 55, 82, 190*). The generation of highly reduced iron-sulfur centers occurs in the Fe protein by the binding of Mg · ATP and in the MoFe protein during the catalytic cycle. The iron of the MoFe protein may also be involved in substrate binding as indicated by shifts in the EPR spectrum in the presence of acetylene (*54*). It has been proposed in chemical models that iron could possibly play a role as a coordination center for substrates.

VII. Reaction Models

Application to and utility of model compounds for the study of nitrogenase have been the subject of several reviews which should be consulted for details (5–7, 184, 216). Chemical models which claim to be relevant to the enzyme must be judged according to the information they give on the biocatalyst.

Yet, one should abstain from an exigent model. Implicit to such a concept, is the fact that a single model cannot be expected to cover all facets of the nitrogenase reaction.

The fundamentals of the approach by chemical models were laid with the achievement of dinitrogen-metal complexes (4, 8) and later on, by their reduction in organic solvents (217, 218). A cyclic reaction has been developed for the titanocene dimer, $[\pi\text{-}Cp)_2 TiN_2 Ti (\pi\text{-}Cp)_2]$, in tetrahydrofuran and sodium naphthalide as reductant. This system has been considered as a possible model of nitrogenase, with bonding of N_2 to a low oxidation state of a metal, reduction thereof via a nitride and subsequent protonation. More elaborate proposals consider "nitridation" at a "dimetal site" of molybdenum (6) or molybdenum and iron (5).

Most of the proposals for the active site of nitrogenase involve a nitrogen molecule bridged between a dimetal site (5, 6, 147, 216). A monometal site, as indicated by *Chatt et al.* (219), may at least be as likely as a dimetal site. Reaction of $[M(N_2)_2(PR_3)_4]$ (M = Mo or W, R = alkyl or aryl) with sulfuric acid in methanol yields one mole of nitrogen gas, the remaining dinitrogen is reduced to ammonia and traces of hydrazine. The yield of ammonia was highest (90%) with the tungsten complex. In this system the metal has to change its oxidation state from zero to six to feed the required number of electrons into the nitrogen molecule. It was suggested that this electron transfer is "triggered" by a 'soft' → 'hard' ligand exchange, the latter possibly being an oxygen ligand (219). One is tempted to see here a possible role of ATP interaction (phosphate) at the dinitrogen reducing-site of the enzyme.

In addition to nitridation, protonation of a N_2-bridged dimetal complex with diimine, HN=NH, and hydrazine, $H_2N - NH_2$, intermediates also appear to be possible nitrogenase models (7). This finds biochemical support from hydrogen-deuterium exchange (133, see Section V. B for its recent mechanistic interpretation) and hydrazine reduction by the *Azotobacter* nitrogenase (135). Further, it has been shown that a dinitrogen complex of bis(pentamethylcyclopentadienyl)zirconium(II) $\{(C_5(CH_3)_5)_2 Zr\}_2(N_2)_3$, yields hydrazine on protonation (220), as well as a benzenediazonium salt with platinum hydride (221), or a $[(C_5H_5)_2 TiR]_2 N_2$ complex of not fully disclosed structure (222). Complexes of diimine have been obtained from ditertiary phosphine complexes, *trans* $[M(N_2)_2(Ph_2PCH_2CH_2PPh_2)_2]$ (M = Mo or W), with hydrogen bromide (223).

A remarkable feature of the known dinitrogen complexes is their end-on bonding mode and the resulting stability thereof (9). Side-on orientation has been claimed for $[\{(PhLi)_3 Ni\}_2 N_2 \cdot 2 Et_2 O]_2$ with the presumed bonding type (224):

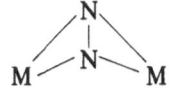

Side-on coordination of dinitrogen was suggested for the titanocene dimer (225) and has recently been demonstrated in a dinitrogen complex of permethyltitanocene (226):

$$M \overset{\diagup^{\displaystyle N}}{\underset{\diagdown_{\displaystyle N}}{}} \vert\vert\vert$$

Side-on bonding to iron and subsequent additional end-on bonding to molybdenum has been proposed for a coupled proton-electron transfer mechanism common to all Mo-enzymes (227). The proposal implies a change in the oxidation state of molybdenum, Mo(IV) of Mo(VI), by donating electrons to the substrate making two Mo-coordinated nitrogen atoms more acidic, which then donate their protons to the substrate. A repetitive mechanism of this type will finally lead to ammonia.

A single molybdenum site with side-on bonding of dinitrogen is the basis for a proposed mechanism of molybdothiol complexes which show many essential features of the enzyme (184, 228). The reactive species is thought to be a monomeric cysteine Mo(IV) complex with two *cis* coordination positions (228):

The initial step of reduction of dinitrogen has been proposed to occur analogous to that of acetylene (184):

$$Mo(IV) + N_2 \rightarrow Mo \overset{\diagup^{\displaystyle N}}{\underset{\diagdown_{\displaystyle N}}{}}\vert\vert \xrightarrow{+\ 2\ H^+} Mo(VI) + N_2H_2$$

The diimine formed disproportionates:

$$3\ N_2H_2 \rightarrow 2\ N_2 + H_2 + N_2H_4$$

Hydrazine subsequently bound in an end-on fashion at the same reduced Mo(IV) site is then further reduced to ammonia. A peculiarity of the Mo-cysteine complex and the Mo-glutathione complex (185) is their enhanced reactivity in the presence of nucleosidetri-phosphates. Phosphorylation of a Mo-bound oxo group has been suggested to generate a leaving group and a vacant coordination site with the appropriate acid-base properties or enhanced reducibility (7, 227, 229, see equation below).

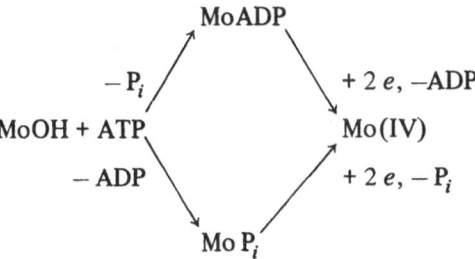

VIII. Ancillary Proteins

Ferredoxins. The mutual interlacing of ferredoxin with nitrogenase on several levels is treated in various contexts in this article. Early research on ferredoxins was summarized by *Buchanan (230)* in volume 1 of this series. Since then, an overwhelming amount of data on ferredoxins and other iron-sulfur proteins has been accumulated. For a complete coverage of this research the reader is referred to a comprehensive treatise of iron-sulfur proteins *(231)*. Both nitrogenase components belong to the class of iron-sulfur proteins, although a relation to ferredoxins awaits further evidence on the structure of the prosthetic groups of nitrogenase. To stress a possible relationship, the two nitrogenase components were named azoferredoxin and molybdoferredoxin *(30)*. Despite early introduction of this nomenclature into the literature, it has not met general acceptance due to the uncertainty of classification. The isolation of ferredoxins with numbers of iron atoms intermediate to spinach and clostridial ferredoxin is still in progress. At present the ferredoxins of *B. polymyxa (232), D. desulfuricans (233), Spirochaeta aurantia (234)*, and *D. gigas (235)* are known to have four iron-sulfur groups. Two types of ferredoxins with 4 Fe ~ S* and 8 Fe ~ S* groups have been found in several organisms *(232, 236, 237)*. To date, only the structures of the prosthetic groups of spinach ferredoxin *(238)*, and *P. aerogenes* ferredoxin *(239)* are well documented. Synthetic analogues for the two Fe and four Fe iron-sulfur proteins have been prepared that show essential properties of the protein-bound groups *(240–243)*. These compounds are of great interest as models for the iron-sulfur groups of nitrogenase and as a possible tool to insert iron-sulfur clusters into the enzyme.

Flavodoxins. These are the smallest known flavoproteins (mol. wt range 15000 to 23000) with FMN as prosthetic group. Their first representative was discovered in a blue-green alga, *Anacystis nidulans (244)* and later its presence was reported independently in *C. pasteurianum (245)*. Flavodoxins can substitute for ferredoxins in many reactions and as a consequence they are synthesized in the cell, particularly under conditions of iron deficiency when ferredoxins may become limited. It was the search for an impaired nitrogenase in bacteria from low-iron media which led to the discovery of bacterial flavodoxin. Although these flavoproteins were first thought to be a peculiarity of prokaryotic organisms, they were also found in green algae, again synthesized there in response to the availability of iron (*246*). Attention to flavodoxins stems primarily from interest in flavin catalysis and the objective to reveal the reaction-priming nature of the protein moiety in a variety of flavin-catalyzed reactions. The elucidation of the primary (*247*) and tertiary structure (*248, 249*) as well as flavin-protein interactions (*250*) of several flavodoxins have been pursued with considerable interest. Flavodoxins exist in three oxidation states; the redox pair semiquinone-fully reduced flavodoxin, with a redox potential of -373 mV at pH 7.0 (*251*), is possibly involved when flavodoxin serves as an electron donor for nitrogenase. A detailed account on flavodoxins can be found elsewhere (*252*).

Hydrogenase. Hydrogenase has no direct relationship to N_2 fixation but is generally a component of N_2-fixing organisms (*11*). An extensive survey has been compiled recently for this enzyme (*253*). Hydrogenase catalyzes the reaction:

$$H_2 + \text{electron acceptor} \rightleftharpoons \text{electron donor} + 2 H^+$$

The reaction is readily reversible and has been used to couple dihydrogen as reductant source to the nitrogenase system via ferredoxin [*e.g.* Ref. (*128*)]. Hydrogen evolution by N_2-fixing organisms is prone to confusion, since it might either be due to classical hydrogenase or due to the nitrogenase-linked system of H_2 evolution. Both systems are distinguishable by their CO reactivity and ATP requirement. Hydrogenase is inhibited by CO (*254, 255*), whereas H_2 evolution related to nitrogenase requires ATP and is CO insensitive (*27, 120, 175, 178*). The situation is less equivocal with *Azotobacter* which has a membrane-bound hydrogenase incapable of H_2 evolution (*256, 257*). Hydrogenases have been obtained in a partially purified state from several organisms, *e.g. D. vulgaris* (*258*) and *C. pasteurianum* (*259*). There has been some controversy as to whether or not the enzyme is a molybdoprotein (*259, 260*) but recent analysis has shown that it is an iron-sulfur protein (*261, 262*). Hydrogenases have a molecular weight between 50000 and 60000 (*258, 261, 263*). Two subunits and a total of four iron-sulfur groups were identified; half of the iron could be removed without affecting the enzymatic activity (*264*). During a study of the redox-state dependence of the EPR signals of clostridial hydrogenase, enzyme preparations were obtained which had more than twice the specific activity of previously reported homogeneous hydrogenase (*261*), but still contained contaminants up to 20%. The EPR spectrum of these preparations was better resolved than previously and more closely resembled the eight Fe iron-sulfur proteins than that of two or four Fe iron-sulfur proteins. Partial oxidation with ferricyanide produced a new complex EPR signal. Further oxidation by ferricyanide or by catalytic amounts of ferredoxin or methyl viologen produced a rhombic signal with approximate *g*-values of 2.09, 2.03, and 1.99. This resonance is atypical for iron-sulfur proteins with both *g*-values and $g_{av} > 2.0$. The intermediate EPR signal was observed between -400 to -380 mV and the rhombic signal showed a 50% decrease at $+300$ mV. These findings were difficult to reconcile with an operative two Fe iron-sulfur structure and suggested either a still heterogeneous preparation and/or the presence of more than two functional iron atoms (*W.G. Zumft* and *G. Palmer*, unpublished). *J.S. Chen*, in parallel efforts, has purified hydrogenase to a specific activity about six-fold higher than that previously obtained. This hydrogenase contained about 12 Fe $\sim$ S* groups and was not composed of subunits (*265*).

Another iron-sulfur protein has been isolated and partially characterized from N_2-fixing extracts of *C. pasteurianum* by *Hardy et al.* (*266*). Due to its characteristic EPR signal, but unknown physiological role, it was named paramagnetic 1.94 protein. It has not been established whether it has a function in N_2 fixation, but the protein is partially associated with clostridial and *Azotobacter* nitrogenases (*42, 267*).

Association of *A. vinelandii* nitrogenase with a hemoprotein has also been noted (*40, 48*). Recently *Bulen et al.* (*268*) crystallized a novel *b*-type cytochrome from *A. vinelandii* which contained an astonishingly high amount of non-heme iron but no acid-labile sulfur. Its possible identity with the nitrogenase-associated cytochrome remains to be seen as well as any participation in N_2 fixation.

Leghemoglobin. N_2-fixing root nodules of legumes contain an O_2-binding hemoprotein resembling myoglobin or the monomeric subunits of hemoglobin. The protein has an extremely high affinity for oxygen, the pO_2 of half oxygenation of leghemoglobin *a* being only 0.04 mmHg. The protein is well characterized, including its primary structure (*269, 270*) and tertiary structure at 5 Å resolution (*271*). Leghemoglobin is localized outside the bacterial cell but within small membranous sacs that envelop groups of bacteroids (*272, 273*). Its concentration therein was estimated to be 1.5 mM (*274*). The biosynthesis of leghemoglobin exhibits an interesting feature in so far as the protein moiety is synthesized by the plant (*275*) and the heme group is contributed by the bacterium (*276*). A possible involvement of this protein in N_2 fixation has always intrigued researchers since its discovery. Proposed functions are O_2 binding, N_2 binding, N_2 reduction via a free radical, hydroxylamine reduction or a role as redox component (*277*). However, the demonstration of N_2 fixation in leghemoglobin-free bacteroids of root nodules precludes an obligatory involvement in N_2 reduction (*103*). At present it appears most convincing that leghemoglobin facilitates the diffusion of dioxygen to the bacteroid surface (*274, 278, 279*). Thus, an efficient oxidative phosphorylation by the bacteroid cell is ensured, yielding high ATP levels which in turn are required for optimal functioning of nitrogenase. This role of leghemoglobin finds impressive support by the replacement of this protein with other hemoglobins or O_2-carrier proteins which have not necessarily been hemoproteins (*279*). It has been suggested that nicotinic acid acts as a regulatory metabolite in modifying the O_2-binding properties of leghemoglobin (*274*).

IX. Generation of Reductant

Nitrogenase requires an electron source of considerably low redox potential which only ferredoxins and flavodoxins are able to provide. No physiological cofactor other than these two proteins has been found. Ferredoxins and flavodoxins are linked via dehydrogenases or oxidoreductases with the principal catabolic lines of energy metabolism. The impressive effectiveness of pyruvate, dihydrogen, or formate to support dinitrogen fixation in fermentative organisms [see Ref. (*280*) for a review], has long overshadowed the possibility that ferredoxins and flavodoxins may be reduced by pyridine nucleotides. *Streicher* and *Valentine* (*281*) state: "The finding of reduced pyridine nucleotide-driven nitrogen fixation in extracts of so many different organisms leads us to conclude that these carriers are among the most important for linking cellular reducing power to nitrogenase". Table 8 presents examples in which reduced pyridine nucleotides are used as a reductant source. Despite these numerous demonstrations *in vitro*, evidence for their physiological role is still scanty. Thermodynamic objections to the reduction of ferredoxin ($E_m = -420\,\mathrm{mV}$) by NAD(P)H ($E_m = -320\,\mathrm{mV}$) have been raised (*282*). To resolve this problem, a high NAD(P)H/-

Table 8. Sequences of electron transfer linked to nitrogenase

Organism	Substrate	Electron transfer			Ref.
A. chroococcum		NADH →	NADH dehydrogenase	→ BV	(*341*)
A. vinelandii	Glucose-6-P Isocitrate Malate α-Ketoglutarate	NADPH →	ferredoxin-NADP reductase	→ Fd + Fld + heat labile factor	(*342*)
A. cylindrica	Glucose-6-P Isocitrate	NADPH →	ferredoxin-NADP reductase (from spinach)	→ Fd	(*292, 293*)
	Pyruvate		Pyruvate-ferredoxin oxidoreductase (phosphoroclastic reaction)	→ Fd	(*290, 291*)
B. polymyxa [a])	Glucose-6-P	NADPH →	ferredoxin-NADP reductase	→ Fd	(*343*)
C. pasteurianum	Pyruvate		Pyruvate-ferredoxin oxidoreductase (phosphoroclastic reaction)	→ Fd or Fld	(*173, 245*)
	Glucose	NADH →	NADH-ferredoxin reductase (acetyl-CoA activated)	→ Fd	(*288*)
K. pneumoniae	Pyruvate Formate Malate		Sequence not established; acceptor Fld		(*344*)
	Glucose-6-P	NADPH →	reductase (crude extract)	→ Fld	(*344*)
Soya-bean bacteroids	Glucose-6-P	NADPH →	reductase (crude extract)	→ Fld + FeS-Protein	(*345*)

[a]) Also pyruvate (*343, 346*) and formate (*334, 343*).
Abbreviations:
Fd, Ferredoxin; Fld, Flavodoxin; BV, Benzyl viologen, FeS-Protein, a partially characterized iron-sulfur protein (*347*).

$NAD(P)^+$ ratio (*183, 280, 283*), a low pH inside the cell, and compartmentation of the reactants to impede free diffusion has been suggested (*283*). The steady-state $NADH/NAD^+$ ratio of the growing *Clostridium* cell has been found to be 0.33 (*283*) with a pH around 6.4 (*284*). In *Azotobacter* the decrease of the $NAD(P)H/NAD(P)^+$ ratio from the maximal to almost the minimal value, due to increased oxygenation of the cell, allowed at the same time a threefold increase in nitrogenase activity (*285*). Hence, neither mechanism appears effective *in vivo* to bring the redox potential of the pyridine nucleotides low enough to ensure significant reduction of ferredoxin. In many cases the driving force *in vitro* was an $NAD(P)H$-generating system to maintain an extreme low concentration of free $NAD(P)^+$. The necessary dehydrogenases or oxidoreductases await, in many instances, identification or fractionation, especially where missing components in the electron transfer chain were supplied by another organism (*e.g.* spinach NADP reductase).

Compartmentation has to be considered as a main factor in ensuring the reduction of ferredoxins or flavodoxins by reduced pyridine nucleotides. The nitrogenase of *A. vinelandii* was found to be associated with a membrane fraction distinct from respiratory membranes (*286*), and *Chromatium* nitrogenase from sonicated cells is apparently particle-bound (*287*). In other N_2-fixing organisms, however, no indications exist for a membrane involvement in the fixation process. With this aspect in mind, it might be worthwhile to heed this possibility more closely. Even without compartmentation by membranes, it is still conceiveable that a highly defined structural arrangement of the components of the nitrogenase-linked electron transfer chain prevents free diffusion of the reactants and results in a vectorial reaction.

If one generally accepts pyridine nucleotides as an alternative reductant source for nitrogenase – of course being of great advantage for the organism not having to depend on a single reductant supply for a vital metabolic process – one would like to know the exact part they contribute to dinitrogen fixation *in vivo*. An evaluation, based on fermentation balances revealed that NADH provides approximately 25% of the required reducing power for N_2 fixation (*288*). This is the only example to date where convincing evidence exists for a contribution of pyridine nucleotides to dinitrogen fixation *in vivo*. Formate and dihydrogen are no longer considered as physiological reductant sources in this organism, since formate synthesis and dihydrogen evolution rather than the reverse processes are the likely function of ferredoxin: CO_2 oxidoreductase (*289*) and hydrogenase (*283*), respectively.

At this point, if one directs attention again for a moment to photosynthetic organisms it appears almost mandatory that reductant and ATP are provided directly by photosynthesis. Although some experimental evidence supports this view, there are also impressive data to the contrary [for a discussion see Ref. (*23*) and references therein]. To summarize the main arguments: (i) inhibitors of photosystem II activity do not immediately inhibit dinitrogen fixation, (ii) there appears to be a general involvement in N_2 fixation of photosystem I only, (iii) blue-green algae also fix dinitrogen to a limited extent in the dark, and (iv) the stoichiometry between N_2 fixation and photosynthesis is variable. This evidence at least requires alternative sources of reductant besides photosynthesis. One of them has now positively been identified. The phosphoroclastic reaction, which was generally assumed to be absent from aerobic organisms, has been demonstrated for the blue-green alga *A. cylindrica* (*290, 291*). It was previously shown for this alga that NADPH was also a possible source of reducing equivalents for nitrogenase (*292, 293*).

The difficulties and uncertainties concerning the reductant source for nitrogenase fortunately have not hindered research on the enzymatic level itself. This is due to the finding by *Bulen et al.* (*120*) that dithionite is directly a reductant for the enzyme. Practically all nitrogenases have been coupled to dithionite since. Dithionite has been shown to undergo an equilibrium:

$$S_2O_4^{2-} \rightleftharpoons 2 SO_2^-$$

Either species acts as a reductant, but SO_2^- appears generally to be the more reactive form (*294, 295*). The formation of SO_2^- can be rate-limiting under certain conditions and should be taken into account in further research.

X. Regulation of Nitrogenase

A great deal of biochemical knowledge about dinitrogen fixation has to be attributed solely to research with *Azotobacter* and *Clostridium*. The forthcoming organism, however, appears to be *Klebsiella*, due to its genetic potential and close relationship to *Escherichia coli*; the latter being considered as the best understood living cell. Though mutants of *A. vinelandii* (*296, 297*) and *C. pasteurianum* (*298*) affected in dinitrogen fixation have been obtained, no genetic transfer or mapping of the genes of dinitrogen fixation (*nif*) has been achieved. Transfer of *nif* genes from *K. pneumoniae* M5A1 to non-fixing (*nif⁻*) mutants of this organism was accomplished in 1971 by *Valentine* and co-workers with the generalized transducing phage P1 of *E. coli* (*299*). Transfer occurred at a frequency of $1-4 \cdot 10^{-5}$ per infective phage. Several transductional crosses between *nif⁻* mutants with transducing frequencies similar to the wild type indicated mutations in rather distant gene loci, however, the majority of crosses gave low frequencies of transductants with location of the mutations near *nif*-95 and *nif*-98 or in between both. Cotransductional analysis of *nif⁻* and *his⁻* double mutants showed a linkage of 78% to the histidine operon (*his*). A chromosome length of 15 to 20 genes was estimated from cotransduction frequencies (*300*), which leaves room not only for the three structural genes required for the three nitrogenase polypeptides but also for a number of regulatory and ancillary *nif* genes which have as yet to be established, *e.g.* genes involved in the metal prosthetic groups. A few *nif⁻* mutants did not map close to the *his* region (*300*) and two mutants were simultaneously affected in nitrate reduction (*299*). From *R. meliloti* a number of pleiotropic mutants involving nitrate reductase and nitrogenase were identified (*301*). These mutants are possibly affected in molybdenum utilization or in a common control mechanism linked to ammonia repression of both enzymes.

Conjugational transfer of the *nif* genes was achieved shortly after gene transduction (*302*). Fertile strains of *K. pneumoniae* do not fix dinitrogen, and in turn the fixing strain M5A1 does not mate with the fertile strains K1–8. However, intrageneric chromosomal transfer occurs after introduction of a de-repressed R-factor from *E. coli* into strain M5A1. *nif⁻* Mutants of M5A1 when mated with the wild type carrying the R-factor, gave recombinant colonies at a frequency of 10^{-5} per donor organism that showed acetylene-reducing activities comparable to those of the wild type. The close linkage to the *his* operon found by *Streicher*, *Gurney* and *Valentine* was confirmed (*302*). The mutants *nif*-9 and *nif*-2 showed a linkage to *his* of 95% and 85%, respectively. Transfer of the *nif* genes was also extended to *E. coli*, being the first example of intergeneric crosses (*303*). With knowledge of the linkage of *nif* genes to the *his* operon, a strain of *Klebsiella* M5A1, carrying the R-factor, was selected for high frequency transfer of *his*. The non-modifying and non-restricting *his⁻* auxotroph strain C of *E. coli* served as recipient. Recombinants with *his⁺* appeared at a frequency of 10^{-7} per donor cell. Three isolated hybrids showed acetylene reduction; the most active one at about half the rate of the wild type. Dinitrogen fixation was not equally correlated to acetylene reduction in these hybrids; either this difference is fortuitous (*303*) or indicates a modification of the nitrogenase in these hybrids. Structural and regulatory *nif* genes were transferred concomitantly to *E. coli*. Acetylene reduction disappeared when the hybrids were grown on fixed nitrogen, thus showing the usual phenomenon of nitrogenase repression. Only one hybrid, C–M7, was stable when subcultured on nutrient broth and allowed a partial characterization of its nitrogenase (*304*). The hybrid C–M7 exhibited oxygen sensitivity under N_2 but not when grown on ammonia. Preliminary evidence indicated the presence of two nitrogenase components, separable by DEAE-cellulose chromatography. The presumed MoFe protein reacted positively with a MoFe protein antiserum of the donor strain, and a catalytically active enzyme hybrid was obtained by cross reaction with the Fe protein of the wild type. Although subtle differences would have escaped detection, essential features of the "nitrogenase from *E. coli*" showed the common nitrogenase characteristics.

The *nif* genes are apparently incorporated into the *E. coli* chromosome. The sedimentation pattern of covalently closed circular DNA of the hybrid C–M7 was indistinguishable from that of the recipient *E. coli* C 603 (with R-factor); only one plasmid DNA, the R-factor, was detectable in both strains (*304*).

Of paramount interest is the exact location of the *nif* genes within the bacterial chromosome. Deletions by the phage P2 had already indicated in earlier work a proximity to the genes of gluconate-6-phosphate dehydrogenase, *gnd* (*300*). A systematic study of deletions yielded three groups of mutants:

(i) *gnd⁻*, *his⁻*, *his* D⁺ (histidinol dehydrogenase), *nif⁺*,
(ii) *gnd⁻*, *his* D⁻, *nif⁻*, *shu⁺* (shikimate utilization), and
(iii) *gnd⁻*, *his* D⁻, *nif⁻*, *shu⁻* (*305*).

The gene order *gnd*, *his* operon, *nif*, *shu* was assumed. Genetic mapping of the deletions by introduction of various F′ *his⁻* episomes indicated a location ot the *nif* genes next to the operator end of the *his* operon. To establish the gene sequence of the *E. coli* hybrid

C—M7, phage EC 1 was used to select for spontaneous segregants (*304*). The analysis of their deletions revealed five groups:

 (i) deletion of the *rfb* locus (loss of phage resistance),
(ii) *rfb⁻*, *nif⁻*,
(iii) *rfb⁻*, *nif⁻* · *gnd⁻*,
(iv) *rfb⁻*, *nif⁻*, *gnd⁻*, *his⁻*, and
 (v) like group (iv) with additional deletions.

This suggests the order *rfb*, *nif*, *gnd*, *his*. This deviation from the gene sequence found in *Klebsiella* has been related to "a complex series of rearrangements of chromosome segments being required for insertion of the heterologous *K. pneumoniae* DNA carrying the *nif* genes into the *E. coli* chromosome" (*306*).

Unlike the *E. coli* hybrid C—M7, two other hybrids, C—M9 and C—L4, carry the *nif* genes on plasmids (*307*). Three plasmids were present in hybrid C—M9 and four in hybrid C—L4. This multiplicity and technical difficulties impeded definite functional assignments for each plasmid. Yet the important significance of incorporating the *nif* genes into plasmids is that they are transferred at a much higher frequency and their transfer is not hindered by genetic incompatibility of host and donor strain. Intrageneric transfer of *nif* genes has also been claimed for the blue-green alga *Nostoc muscorum* (*308*). The regulation of gene expression in N_2-fixing symbiotic *Rhizobia* has always been an intriguing problem. *Dunican* and *Tierney* (*309*) introduced an R-factor into *R. trifolii* strain T1K and succeeded in *nif* transfer to non-fixing *K. aerogenes* 418. Though these authors were aware of possible gene expression of latent *nif* genes in the recipient, this is an unlikely alternative in the light of evidence with other R-factor-mediated gene transfers. The *nif* expression in *K. aerogenes* thus shows that *Rhizobium* must bear all the necessary genetic information for dinitrogen fixation, although the conditions for its transcription remain to be specified. Recently it was demonstrated with a serological technique that free-living and consequently non-fixing *R. japonicum* synthesized an inactive MoFe protein, or part of it, when grown anaerobically with nitrate or nitrate and ammonia (*310*). Repeated attempts to characterize the DNA of free-living *Rhizobia* and that of bacteroids remain equivocal (*311, 312*). The discovery of a plasmid-like DNA in bacteroids could be of potential importance (*311*). The question whether *Rhizobium* harbors all the genes of dinitrogen fixation and does not require genetic material from its host plant will definitively be answered when N_2-fixing mutants (a fruitless undertaking so far) are obtained. Success in this direction now appears to be forthcoming.

The nitrogenase genes are presumably organized as one or several *nif* operons, adjacent to the *his* operon (*300, 304*). The existence of an operon is suggested by several lines of evidence. The effect of ammonia repression on nitrogen is exhibited both by the *E. coli* hybrids (*303*) and the *Klebsiella* hybrids with *Rhizobium* as *nif* donor (*309*). Repression and derepression of the Fe protein and the MoFe protein in *A. vinelandii* (*126*) and *A. chroococcum* (*99*) are coordinate. Both nitrogenase components are synthesized in the presence of tungsten, although the MoFe protein is catalytically inactive (*197*). At least two operons appear to be required for nitrogenase from the observation that the *A. vine-*

landii mutant UW38 produces the Fe protein in large excess, but does not synthesize the MoFe protein (*313*). On the other hand, revertants which fix dinitrogen exhibit low activities in both components, *i.e.* do not continue to hyper-produce the Fe protein. This suggests rather a single mutation in a common regulatory site. *Shah et al.* (*313*) proposed a model with two separate operons for the MoFe protein and the Fe protein, both of which are regulated by the same control gene. The latter is believed to produce an activator protein that exists in two states activating specifically either the operon of the Fe protein or that of the MoFe protein. Two nitrogenase operons in *Azotobacter* are indicated by the exclusive synthesis of the Fe protein under conditions of molybdenum starvation (*197*), but regulatory phenomena related to the metal prosthetic groups seem dissimilar in different N_2-fixing organisms, since neither *Klebsiella* (*203*) nor *Clostridium* (*314*) synthesize any one of the nitrogenase components in the absence of Mo or in the presence of W. An interesting finding is the non-coordinate synthesis of the nitrogenase components in *Clostridium*, where the appearance of the MoFe protein precedes that of the Fe protein by about 20 min (*315*). The regulation of nitrogenase is extremely complex due to a multifactorial interplay of several variables. Molybdenum might be of importance as an inducer of the enzyme (*197, 203*) but apparently dinitrogen is not (*124*).

The synthesis of nitrogenase in N_2-fixing cultures stops immediately upon addition of ammonia. This effect is well documented although its mechanism remained enigmatic. Nitrogenase itself is not inhibited by ammonia and continues to function in the cell after addition of ammonia, yet is consecutively lost by dilution or "wash-out" upon further growth of the culture (*57, 126, 316*). In addition to this, a short-term influence of ammonia, affecting the active center of the enzyme, might be operative in *A. vinelandii* (*80*). Loss of nitrogenase activity (followed as acetylene reduction, N_2 fixation, and dithionite oxidation) occurred at the same rate as the EPR signal of the MoFe protein declined. However, serologically cross-reacting material of the MoFe protein disappeared at a substantially slower rate. Unfortunately it has not been investigated whether this remaining portion of the enzyme could be reactivated rapidly when ammonia is replaced by dinitrogen. Ammonia-grown cells, switched to dinitrogen, exhibit a lag period of variable length without nitrogenase activity and with stagnant growth (*99, 126, 316*). It has been shown that this lag period is dependent on the growth conditions and nutritional status of the cell (*57, 316*). Specific inhibitors of RNA transcription and translation indicate that ammonia exerts its effect at the level of mRNA synthesis (*57*). The decay of the coding capacity after initiation of derepression has a half life close to 5 min (*57*). Results similar to those with *Klebsiella* have been reported for *Clostridium* (*315*).

Glutamine (*124*), asparagine (*317*), urea (*129*), and nitrate (*99*) all repress nitrogenase. Ammonia has been generally assumed to be an effector common to these compounds. The glutamate analogue, methionine sulfone and methionine sulfoximine, inhibit the glutamate and glutamine synthetases but not the glutamate dehydrogenase of *K. pneumoniae* (*318*). A similar inhibition pattern is obtained with *A. vinelandii*, but methionine sulfoximine inhibits glutamate synthase only slightly. However, under the action of these inhibitors both organisms continue to synthesize nitrogenase in the presence of ammonia, which suggests that a secondary mechanism is involved rather than NH_4^+ being the actual effector of nitrogenase repression. Cyclic AMP might be involved in the expression of this secondary mechanism. Contrary to a previous report (*99*) preliminary evidence has now been found that cAMP stimulated nitrogenase synthesis during derepression (*319*).

Many, if not all, of these mosaic findings may be related to a mechanism that involves glutamine synthetase in the regulation of nitrogenase biosynthesis (*320, 321*). The intricate regulatory system of histidine utilization (*hut* operon) has been shown to be linked to the level of glutamine synthetase (*322*); the latter in its deadenylylated form being an activator of the transcription of the *hut* operon (*323*). A similar mechanism is now envisaged to regulate the *nif* operon(s). Glutamine auxotrophs of *K. pneumoniae* that did not synthesize catalytically active glutamine synthetase also lacked nitrogenase. When the missing genes of glutamine synthetase were introduced into these mutants by an *E. coli* episome F′ 133 carrying them, glutamine synthetase and nitrogenase activity were restored equally. Another class of *K. pneumoniae* mutants with a constitutive glutamine synthetase produced nitrogenase in the presence of NH_4^+ up to 30% of the derepressed level (*320, 321*). These mutants resemble those of *A. vinelandii* which synthesize nitrogenase in the presence of NH_4^+ at 25% of its derepressed level (*324*). An investigation of the glutamine synthetase level in these mutants is awaited with considerable interest to see whether the proposed mechanism generally occurs. Finally, introduction of an F′ *nif-his* factor, carrying the *nif* genes of *K. pneumoniae* M5A1, into non-fixing *Klebsiella* mutants with constitutive‾ glutamine synthetase, yielded hybrids which showed nitrogenase activity around 10–20% of its derepressed level (*321*). These findings represent strong evidence for a regulatory role of glutamine synthetase in nitrogenase biosynthesis, similar to that of the *hut* operon.

Control of the enzyme at a genetic level must have an effective counterpart which modulates the activity of the enzyme itself. Electron flow and the available ATP (the energy charge will certainly provide a better measure than the ATP concentration alone) con be envisaged to exert a regulatory effect on nitrogenase. Since these modulators are interlaced by complex reaction sequences, their functioning *in vivo* will be a challenge for future research. In certain organisms a third factor, dioxygen, must be taken into account. Dioxygen is related in aerobic organisms to the energy supply, as well as in anaerobic and facultative anaerobic organisms to direct effects on the enzyme itself. Several points concerning dioxygen and nitrogenase have been considered in Section IV.D.

XI. Conclusions

The past decade has brought many new insights into the biological reduction of dinitrogen and its biocatalyst, nitrogenase. We now have a detailed knowledge of most of the biochemical properties of the enzyme, which allows us to conclude an overall similarity of nitrogenases from organizationally different organisms. At present, little information about the enzyme of photosynthetically N_2-fixing organisms is available, yet as far as the data allow a generalization, this enzyme fits the pattern derived from non-photosynthetic bacteria. Apparently only one nitrogenase has evolved and has been preserved in pro-karyotic organisms. The application of low-temperature EPR and Mössbauer spectroscopy allowed a first probing of the operating enzyme at the molecular level. The results will have to be incorporated in any future formulation of the reaction mechanism. Many generalized schemes of the nitrogenase reaction have been brought forward. I shall not attempt to further proliferate their number by still another speculative formulation.

A few years ago *Hardy et al.* (*147*) compiled a list of unsolved problems in dinitrogen fixation. Since then progress has been made in many areas including some of the central questions. Many discrepancies remain, however, and rather than bias one-sided conclusions, these incoherencies have been stated as such. Unfortunately the field of dinitrogen fixation is plagued by artifacts due to the extreme difficulty in handling this enzyme. The mastery of anaerobic techniques by a large number of biochemists will certainly improve this situation in the near future.

At present, the elucidation of the structure of the prosthetic groups of both nitro-genase components is of eminent interest. Furthermore, there is a strong need to clarifiy the role of molybdenum in enzyme catalysis. The characterization of the molybdenum cofactor will probably give a completely new insight into this area. Finally, there is a tremendous task ahead to unravel the mechanism of ATP action. Interlacing with other energy transducing processes can be envisaged, especially on the background of the now clearly recognized participation of iron-sulfur proteins in energy transduction.

Genetic research doubtlessly will have a great impact on the eventual bioengineering of new N_2-fixing systems. Even with full understanding of the role of metals in the enzy-matic mechanism, and with development of catalysts as a result of this knowledge, the costs of dihydrogen in any industrial process of ammonia synthesis will not have been overcome. Hope to satiate the immediate need for fixed nitrogen thus centres around the proven biological potential of self-sustained or symbiotic dinitrogen fixation. It appears feasible to spread this unique property to other bacteria or even eukaryotic systems via phages, extracted DNA or plasmids (*306, 325, 326*). Other approaches are more directly oriented at the symbiotic system of *Rhizobia* with legumes and investigate its propagation to other plants (*327, 328, 329*). Finally, excretion of ammonia by genetic manipulation of N_2-fixing bacteria seems to be another possible way to increase the supply of fixed nitrogen under abolition or at least curtailment of the use of nitrogen fertilizers (*330*).

Acknowledgements. I wish to thank Drs. *C.A. Appleby, W.A. Bulen, R.H. Burris,* and *J.S. Chen* for providing manuscripts prior to publication. I am indebted to *Janet Westpheling* for her efficient help in preparing the manuscript. Contributions of the author were supported by the Deutsche Forschungs-gemeinschaft.

References

1. *Silver, W. S., Postgate, J. R.:* J. theoret. Biol. *40*, 1 (1973).
2. *Wilson, P. W.:* in Hdb. der Pflanzenphysiol. (Ruhland, W., ed.) vol. *8*, pp. 9. Berlin: Springer-Verlag (1958).
3. *Wilson, P. W.:* Proc. Roy. Soc. London, Ser. B, *172*, 319 (1969).
4. *Allen, A. D., Senoff, C. V.:* Chem. Commun. 621 (1965).
5. *Leigh, G. J.:* The Chemistry and Biochemistry of Nitrogen Fixation (Postgate, J. R., ed.), pp. 19. London: Plenum Press (1971).
6. *Chatt, J., Richards, R. L.:* The Chemistry and Biochemistry of Nitrogen Fixation (Postgate, J. R., ed.), pp. 57. London: Plenum Press (1971).
7. *Hardy, R. W. F., Burns, R. C., Parshall, G. W.:* in Inorganic Biochemistry (Eichhorn, G. L., ed.), vol. 2, pp. 745. Amsterdam: Elsevier (1973).
8. *Allen, A. D., Harris, R. O., Loescher, B. R., Stevens, J. R., Whiteley, R. N.:* Chem. Rev., *12*, 11 (1973).
9. *Sellmann, D.:* Angew. Chem. Int. Ed. *13*, 639 (1974).
10. *Burk, D.:* Ergeb. Enzymforsch., *3*, 23 (1934).
11. *Burris, R. H.:* in The Chemistry and Biochemistry of Nitrogen Fixation (Postgate, J. R., ed.), pp. 105. London: Plenum Press (1971).
12. *Hardy, R. W. F.:* Treatise on Dinitrogen (N_2) Fixation. New York: Wiley-Intersience (1975), in press.
13. *Hardy, R. W. F., Holsten, R. D., Jackson, E. K., Burns, R. C.:* Plant Physiol., *43*, 1185 (1968).
14. *Burris, R. H., Wilson, P. W.:* Meth. Enzymology, *4*, 355 (1957).
15. *Conway, E. J.:* Microdiffusion Analysis and Volumetric Error, Crosby Lockwood, London (1957).
16. *Moore, S., Stein, W. H.:* J. Biol. Chem. *211*, 907 (1954).
17. *Dilworth, M. J., Subramanian, D., Munson, T. O., Burris, R. H.:* Biochim. Biophys. Acta *99*, 486 (1965).
18. *Umbreit, W. W., Burris, R. H., Stauffer, J. W.:* Manometric Techniques. Minneapolis, Minn.: Burgess Publ. Co. (1964).
19. *Taussky, H. H., Shorr, E.:* J. Biol. Chem. *202*, 675 (1953).
20. *Bartlett, G. R.:* J. Biol. Chem. *234*, 466 (1959).
21. *Eggleton, P., Elsden, S. R., Gough, N.:* Biochem. J. *37*, 526 (1943).
22. *Millbank, J. W.:* Arch. Mikrobiol. *72*, 375 (1970).
23. *Stewart, W. D. P.:* Annu. Rev. Microbiol. *27*, 283 (1973).
24. *Carnahan, J. E., Mortenson, L. E., Mower, H. F., Castle, J. E.:* Biochim. Biophys. Acta *44*, 520 (1960).
25. *Mortenson, L. E.:* Biochim. Biophys. Acta *81*, 473 (1964).
26. *McNary, J. E., Burris, R. H.:* J. Bacteriol. *84*, 598 (1962).
27. *Bulen, W. A., LeComte, J. R.:* Proc. Nat. Acad. Sci. USA *56*, 979 (1966).
28. *Kelly, M.:* Biochim. Biophys. Acta *171*, 9 (1969).
29. *Detroy, R. W., Witz, D. F., Parejko, R. A., Wilson, P. W.:* Proc. Nat. Acad. Sci. USA *61*, 537 (1968).
30. *Mortenson, L. E., Morris, J. A., Jeng, D. Y.:* Biochim. Biophys. Acta *141*, 516 (1967).
31. *Biggins, D. R., Kelly, M., Postgate, J. R.:* Eur. J. Biochem. *20*, 140 (1971).
32. *Klucas, R. V., Koch, B., Russell, S., Evans, H. J.:* Plant Physiol. *43*, 1906 (1968).
33. *Whiting, M. J., Dilworth, M. J.:* Biochim. Biophys. Acta *371*, 337 (1974).
34. *Smith, R. V., Telfer, A., Evans, M. C. W.:* J. Bacteriol. *107*, 574 (1971).
35. *Evans, M. C. W., Telfer, A., Smith, R. V.:* Biochim. Biophys. Acta *310*, 344 (1973).
36. *Hwang, J. C., Burris, R. H.:* Biochim. Biophys. Acta *283*, 339 (1972).
37. *Hadfield, K. L., Bulen, W. A.:* Biochemistry *8*, 5103 (1969).
38. *Silverstein, R., Bulen, W. A.:* Biochemistry *9*, 3809 (1970).
39. *Burns, R. C., Holsten, R. D., Hardy, R. W. F.:* Biochem. Biophys. Res. Commun. *39*, 90 (1970).
40. *Shah, V. K., Brill, W. J.:* Biochim. Biophys. Acta *305*, 445 (1973).
41. *Kleiner, D., Chen, C. H.:* Arch. Microbiol. *98*, 93 (1974).

42. *Zumft, W. G., Mortenson, L. E.:* Eur. J. Biochem. *35*, 401 (1973).
43. *Nakos, G., Mortenson, L. E.:* Biochemistry *10*, 455 (1971).
44. *Eady, R. R., Smith, B. E., Cook, K. A., Postgate, J. R.:* Biochem. J. *128*, 655 (1972).
45. *Israel, W. D., Howard, R. L., Evans, H. J., Russell, S. A.:* J. Biol. Chem. *249*, 500 (1973).
46. *Kelly, M., Klucas, R. V., Burris, R. H.:* Biochem. J. *105*, 3c (1967).
47. *Thorneley, R. N. F.:* Biochem. J. *145*, 391 (1975).
48. *Burns, R. C., Hardy, R. W. F.:* Meth. Enzymology 24, 480 (1972).
49. *Mortenson, L. E.:* Meth. Enzymology 24, 446 (1972).
50. *Moustafa, E., Mortenson, L. E.:* Biochim. Biophys. Acta *172*, 106 (1969).
51. *Tso, M.-Y. W., Ljones, T., Burris, R. H.:* Biochim. Biophys. Acta *267*, 600 (1972).
52. *Moustafa, E.:* Biochim. Biophys. Acta *206*, 178 (1970).
53. *Palmer, G., Multani, J. S., Cretney, W. C., Zumft, W. G., Mortenson, L. E.:* Arch. Biochem. Biophys. *153*, 325 (1972).
54. *Smith, B. E., Lowe, D. J., Bray, R. C.:* Biochem J. *135*, 331 (1973).
55. *Orme-Johnson, W. H., Hamilton, W. D., Ljones, T., Tso, M.-Y. W., Burris, R. H., Shah, V. K., Brill, W. J.:* Proc. Nat. Acad. Sci. USA *69*, 3142 (1972).
56. *Zumft, W. G., Mortenson, L. E., Palmer, G.:* Eur. J. Biochem. *46*, 525 (1974).
57. *Tubb, R. S., Postgate, J. R.:* J. Gen. Microbiol. *79*, 103 (1973).
58. *Huang, T. C., Zumft, W. G., Mortenson, L. E.:* J. Bacteriol. *113*, 884 (1973).
59. *Huang, T. C.:* Thesis, Purdue University (1973).
60. *Cárdenas, J., Mortenson, L. E.:* Anal. Biochem. *60*, 372 (1974).
61. *Kelly, M., Lang, G.:* Biochim. Biophys. Acta *223*, 86 (1970).
62. *Eady, R. R.:* Internat. Symp. N_2 Fixation, Washington State University Press, Pullman, in press.
63. *Bray, R. C., Swann, J. C.:* Struct. Bonding *11*, 107 (1972).
64. *Dalton, H., Morris, J. A., Ward, M. A., Mortenson, L. E.:* Biochemistry *10*, 2066 (1971).
65. *Kaijyama, S., Matsuki, T., Nosoh, Y.:* Biochem. Biophys. Res. Commun. *37*, 711 (1969).
66. *Chen, J. S., Multani, J. S., Mortenson, L. E.:* Biochim. Biophys. Acta *310*, 51 (1973).
67. *Dilworth, M. J.:* Annu. Rev. Plant Physiol. *25*, 81 (1974).
68. *Kelly, M.:* Biochim. Biophys. Acta *191*, 527 (1969).
69. *Dalton, H., Mortenson, L. E.:* Bacteriol. Rev. *36*, 231 (1972).
70. *Nakos, G., Mortenson, L. E.:* Biochim. Biophys. Acta *229*, 431 (1971).
71. *Bergersen, F. J., Turner, G. L.:* Biochim. Biophys. Acta *214*, 28 (1970).
72. *Hardy, R. W. F., Burns, R. C.:* in Iron-Sulfur Proteins (Lovenberg, W., ed.) vol. 1, pp. 65. New York: Academic Press (1973).
73. *Stasny, J. T., Burns, R. C., Kornat, B. D., Hardy, R. W. F.:* J. Cell Biol. *60*, 311 (1974).
74. *Morino, Y., Snell, E. E.:* J. Biol. Chem. *242*, 5591 (1967).
75. *Valentine, R. C., Wrigley, N. G., Scrutton, M. C., Irias, J. J., Utter, M. F.:* Biochemistry *5*, 3111 (1966).
76. *Ljones, T.:* Biochim. Biophys. Acta *321*, 103 (1973).
77. *Walker, M., Mortenson, L. E.:* Biochem. Biophys. Res. Commun. *54*, 669 (1973).
78. *Jeng, D. Y., Morris, J. A., Mortenson, L. E.:* J. Biol. Chem. *245*, 2809 (1970).
79. *Zumft, W. G., Palmer, G., Mortenson, L. E.:* Biochim. Biophys. Acta *292*, 413 (1973).
80. *Davis, L. C., Shah, V. K., Brill, W. J., Orme-Johnson, W. H.:* Biochim. Biophys. Acta *256*, 512 (1972).
81. *Smith, B. E., Lowe, D. J., Bray, R. C.:* Biochem. J. *130*, 641 (1972).
82. *Smith, B. E., Lang, G.:* Biochem. J. *137*, 169 (1974).
83. *Carter, C. W. Jr., Freer, S. T., Xuong, Ng. H., Alden, R. A., Kraut, J.:* Cold Spring Harbor Symp. Quant. Biol. *36*, 381 (1971).
84. *Keresztes-Nagy, S., Margoliash, E.:* J. Biol. Chem. *241*, 5595 (1966).
85. *Lovenberg, W., Buchanan, B. B., Rabinowitz, J. C.:* J. Biol. Chem. *238*, 3899 (1963).
86. *Malkin, R., Rabinowitz, J. C.:* Biochemistry, *6*, 3880 (1967).
87. *Petering, D., Fee, J. A., Palmer, G.:* J. Biol. Chem. *246*, 643 (1971).
88. *Bergersen, F. J., Turner, G. L.:* Biochem. J. *131*, 61 (1973).

89. *Wong, P.P., Burris, R.H.:* Proc. Natl. Acad. Sci. USA *69*, 672 (1972).
90. *Yates, M.G.:* Eur. J. Biochem. *29*, 386 (1972).
91. *Benemann, J.R., Smith, G.M., Kostel, P.J., McKenna, C.E.:* FEBS Lett. *29*, 219 (1973).
92. *Dalton, H., Postgate, J.R.:* J. Gen. Microbiol. *54*, 463 (1969).
93. *Phillips, D.H., Johnson, M.J.:* J. Biochem. Microbiol. Technol. Eng. *3*, 277 (1961).
94. *Oppenheim, J., Marcus, L.:* J. Bacteriol. 286 (1970).
95. *Marcus, L., Kaneshiro, T.:* Biochim. Biophys. Acta *288*, 296 (1972).
96. *Swank, R.T., Burris, R.H.:* Biochim. Biophys. Acta *180*, 473 (1969).
97. *Knowles, C.J., Redfarn, E.R.:* Biochim. Biophys. Acta *162*, 348 (1968).
98. *Pate, J.L., Shah, V.K., Brill, W.J.:* J. Bacteriol. *114*, 1346 (1973).
99. *Drozd, J.W., Tubb, R.S., Postgate, J.R.:* J. Gen. Microbiol. *73*, 221 (1972).
100. *Dua, R.D., Burris, R.H.:* Proc. Nat. Acad. Sci. USA *50*, 169 (1963).
101. *Moustafa, E., Mortenson, L.E.:* Anal. Biochem. *24*, 226 (1968).
102. *Haystead, A., Robinson, R., Stewart, W.D.P.:* Arch. Mikrobiol. *74*, 235 (1970).
103. *Koch, B., Evans, H.J., Russell, S.:* Proc. Natl. Acad. Sci. USA *58*, 1343 (1967).
104. *Bergersen, F.J., Turner, G.L.:* J. Gen. Microbiol. *53*, 205 (1968).
105. *Evans, H.J., Koch, B., Klucas, R.:* Meth. in Enzymology *24*, 470 (1972).
106. *Scheraga, H.A., Nemethy, G., Steinberg, I.Z.:* J. Biol. Chem. *237*, 2506 (1962).
107. *Raijman, L., Grisolia, S.:* Biochem. Biophys. Res. Commun. *4*, 262 (1961).
108. *Penefsky, H.S., Warner, R.C.:* J. Biol. Chem. *240*, 4694 (1965).
109. *Irias, J.J., Olmsted, M.R., Utter, M.F.:* Biochemistry *8*, 5136 (1969).
110. *Havir, E.A., Tamir, H., Ratner, S., Warner, R.C.:* J. Biol. Chem. *240*, 3079 (1965).
111. *Kirkman, H.N., Hendrickson, E.M.:* J. Biol. Chem. *237*, 2371 (1962).
112. *Kono, N., Uyeda, K.:* J. Biol. Chem. *248*, 8603 (1973).
113. *Burris, R.H.:* Proc. Roy. Soc. Ser. B. *172*, 339 (1969).
114. *Vandecasteele, J.P., Burris, R.H.:* J. Bacteriol. *101*, 794 (1970).
115. *Mortenson, L.E., Zumft, W.G., Huang, T.C., Palmer, G.:* Biochem. Soc. Trans. *1*, 35 (1973).
116. *Zumft, W.G., Cretney, W.C., Huang, T.C., Mortenson, L.E., Palmer, G.:* Biochem. Biophys. Res. Common. *48*, 1525 (1972).
117. *Eady, R.R.:* Biochem. J. *135*, 531 (1973).
118. *Eady, R.R., Smith, B.E., Thornley, R.N.F., Ware, D.A., Postgate, J.R.:* Biochem. Soc. Trans. *1*, 37 (1973).
119. *Ljones, T., Burris, R.H.:* Biochim. Biophys. Acta *275*, 93 (1972).
120. *Bulen, W.A., Burns, R.C., LeComte, J.R.:* Proc. Natl. Acad. Sci. USA *53*, 532 (1965).
121. *Wilson, P.W., Burris, R.H.:* Annu. Rev. Microbiol. *7*, 415 (1953).
122. *Pengra, R.M., Wilson, P.W.:* J. Bacteriol. *75*, 21 (1958).
123. *Strandberg, G.W., Wilson, P.W.:* Can. J. Microbiol. *14*, 25 (1968).
124. *Parejko, R.A., Wilson, P.W.:* Can. J. Microbiol. *16*, 681 (1970).
125. *Schick, H.-J.:* Arch. Mikrobiol. *75*, 89 (1971).
126. *Shah, V.K., Davis, L.C., Brill, W.J.:* Biochim. Biophys. Acta *256*, 598 (1972).
127. *Schneider, K.C., Bradbeer, C., Singh, R.N., Wang, L.C., Wilson, P.W.:* Proc. Natl. Acad. Sci. USA *46*, 726 (1960).
128. *Mortenson, L.E.:* Proc. Nat. Acad. Sci. USA *52*, 272 (1964).
129. *Hardy, R.W.F., Knight, E. Jr.:* Biochim. Biophys. Acta *139*, 69 (1967).
130. *Strandberg, G.W., Wilson, P.W.:* Proc. Nat. Acad. Sci. USA *58*, 1404 (1967).
131. *Kelly, M.:* Biochem. J. *107*, 1 (1968).
132. *Parker, C.A., Scutt, P.B.:* Biochim. Biophys. Acta *38*, 230 (1960).
133. *Hoch, G.E., Schneider, K.C., Burris, R.H.:* Biochim. Biophys. Acta *37*, 273 (1960).
134. *Jackson, E.K., Parshall, G.W., Hardy, R.W.F.:* J. Biol. Chem. *19*, 4952 (1968).
135. *Bulen, W.A.:* Proc. Internat. Symp. N_2 Fixation, Washington State University Press, Pullman (1975) in press.
136. *Schöllhorn, R., Burris, R.H.:* Proc. Natl. Acad. Sci. USA *57*, 1317 (1967).
137. *Hardy, R.W.F., Knight, E. Jr.:* Biochem. Biophys. Res. Commun. *23*, 409 (1966).

138. *Lockshin, A., Burris, R.H.:* Biochim. Biophys. Acta *111*, 1 (1965).
139. *Dilworth, M.J.:* Biochim. Biophys. Acta *127*, 285 (1966).
140. *Hardy, R.W.F., Jackson, E.K.:* Fed. Proc. *26*, 725 (1967).
141. *Koch, B., Evans, H.J.:* Plant Physiol. *41*, 1748 (1966).
142. *Bergersen, F.J.:* Aust. J. Biol. Sci. *16*, 669 (1963).
143. *Turner, G.L., Bergersen, F.J.:* Biochem. J. *115*, 529 (1969).
144. *Kelly, M.:* Biochem. J. *109*, 322 (1968).
145. *Burns, R.C., Bulen, W.A.:* Biochim. Biophys. Acta *105*, 437 (1965).
146. *Kelly, M., Postgate, J.R., Richards, R.L.:* Biochem. J. *102*, 1c (1967).
147. *Hardy, R.W.F., Burns, R.C., Parshall, G.W.:* Adv. in Chemistry Ser. *100*, 219 (1971).
148. *Bulen, W.A., LeComte, J.R., Burns, R.C., Hinkson, J.:* in Non-Heme Iron Proteins (San Pietro, A., ed.) pp 261–274. Yellow Springs, Ohio: The Antioch Press (1965).
149. *Fuchsman, W.H., Hardy, R.W.F.:* Bioinorganic Chem. *1*, 195 (1972).
150. *Hwang, J.C., Chen, C.H., Burris, R.H.:* Biochim. Biophys. Acta *292*, 256 (1973).
151. *Hardy, R.W.F., Knight, E.Jr., D'Eustachio, A.J.:* Biochem. Biophys. Res. Commun. *20*, 539 (1965).
152. *McKenna, C.E., Benemann, J.R., Traylor, T.G.:* Biochem. Biophys. Res. Commun. *41*, 1501 (1970).
153. *Burns, R.C., Fuchsman, W.H., Hardy, R.W.F.:* Biochem. Biophys. Res. Commun. *42*, 353 (1971).
154. *Burris, R.H., Orme-Johnson, W.B.:* Proc. Internat. Symp. N_2 Fixation, Washington State University Press, Pullman (1975) in press.
155. *Evans, M.C.W., Telfer, A., Cammack, R., Smith, R.V.:* FEBS Lett. *15*, 317 (1971).
156. *Bui, P.T., Mortenson, L.E.:* Proc. Nat. Acad. Sci. USA *61*, 1021 (1968).
157. *Hardy, R.W.F., Knight, E. Jr.:* Biochim. Biophys. Acta *132*, 520 (1966).
158. *Kennedy, I.R., Morris, J.A., Mortenson, L.E.:* Biochim. Biophys. Acta *153*, 777 (1968).
159. *Moustafa, E., Mortenson, L.E.:* Nature (London) *216*, 1241 (1967).
160. *Burns, R.C.:* Biochim. Biophys. Acta *171*, 253 (1969).
161. *Parejko, R.A., Wilson, P.W.:* Proc. Nat. Acad. Sci. USA *68*, 2016 (1971).
162. *Biggins, D.R., Kelly, M.:* Biochim. Biophys. Acta *205*, 288 (1970).
163. *Tso, M.Y.W., Burris, R.H.:* Biochim. Biophys. Acta *309*, 263 (1973).
164. *Thorneley, R.N.F., Eady, R.R.:* Biochem. J. *133*, 405 (1973).
165. *Walker, G.A., Mortenson, L.E.:* Biochem. Biophys. Res. Commun. *53*, 904 (1973).
166. *Walker, G.A., Mortenson, L.E.:* Biochemistry *13*, 2382 (1974).
167. *Buisson, D.H., Sigel, H.:* Biochim. Biophys. Acta *343*, 45 (1974).
168. *Glassman, T.A., Suchy, J., Cooper, C.:* Biochemistry *12*, 2430 (1973).
169. *Wee, V., Feldman, I., Rose, P., Gross, S.:* J. Am. Chem. Soc. *96*, 103 (1974).
170. *Glassman, T.A., Cooper, C., Kuntz, C.P.P., Swift, T.J.:* FEBS Lett. *39*, 73 (1974).
171. *Thorneley, R.N.F., Wilison, K.R.:* Biochem. J. *139*, 211 (1974).
172. *Thorneley, R.N.F.,* Biochim. Biophys. Acta *358*, 247 (1974).
173. *Hardy, R.W.F., D'Eustachio, A.J.:* Biochem. Biophys. Res. Commun. *15*, 314 (1964).
174. *Bulen, W.A., Burns, R.C., LeComte, J.R.:* Biochem. Biophys. Res. Commun. *17*, 265 (1964).
175. *Burns, R.C.:* in Non-Heme Iron Proteins: Role in Energy Conversion (San Pietro, A., ed.), pp. 289. Yellow Springs, Ohio: Antioch Press (1965).
176. *Winter, H.C., Burris, R.H.:* J. Biol. Chem. *243*, 940 (1968).
177. *Bui, P.T., Mortenson, L.E.:* Biochemistry *8*, 2462 (1969).
178. *Mortenson, L.E.:* Biochim. Biophys. Acta *127*, 18 (1966).
179. *Daesch, G., Mortenson, L.E.:* J. Bacteriol. *96*, 346 (1967).
180. *Postgate, J.R.:* in The Chemistry and Biochemistry of Nitrogen Fixation (Postgate, J.R., ed.), pp. 178. London: Plenum Press (1971).
181. *Dalton, H., Postgate, J.R.:* J. Gen. Microbiol. *56*, 307 (1969).
182. *Kennedy, I.R.:* Biochim. Biophys. Acta *222*, 135 (1970).
183. *Mortenson, L.E.:* Survey of Progress in Chemistry, Vol. 4, pp. 127. New York: Academic Press (1968).

184. *Schrauzer, G.N., Kiefer, G.W., Tano, K., Doemeny, P.A.:* J. Am. Chem. Soc. *96*, 641 (1974).
185. *Werner, D., Russell, S.A., Evans, H.J.:* Proc. Nat. Acad. Sci. USA *70*, 339 (1973).
186. *Albrecht, S.L., Evans, M.C.W.:* Biochem. Biophys. Res. Commun. *55*, 1009 (1973).
187. *Walker, M.L., Walker, G.A., Mortenson, L.E.:* Proc. Internat. Symp. N_2 Fixation, Washington State University Press, Pullman (1975), in press.
188. *Watt, G., Bulen, W.A.:* Proc. Internat. Symp. N_2 Fixation, Washington State University Press, Pullman (1975), in press.
189. *Evans, M.C.W., Albrecht, S.L.:* Biochem. Biophys. Res. Commun. *61*, 1187 (1975).
190. *Mortenson, L.E., Zumft, W.G., Palmer, G.:* Biochim. Biophys. Acta *292*, 422 (1973).
191. *Carter, C.W., Jr., Kraut, J., Freer, S.T., Alden, R.A., Siekers, L.C., Adman, E., Jensen, L.E.:* Proc. Nat. Acad. Sci. USA *69*, 3526 (1972).
192. *Eisenstein, K.K., Wang, J.H.:* J. Biol. Chem. *244*, 1720 (1969).
193. *Cammack, R.:* Biochem. Biophys. Res. Commun. *54*, 548 (1973).
194. *Beinert, H.:* in Biological Applications of Electron Spin Resonance (Swartz, H.M., Bolton, J.R., Borg, D.C., eds.) pp. 351. New York: Wiley Interscience (1972).
195. *Bortels, H.:* Zbl. Bakteriol. *87*, 476 (1933).
196. *Fay, P., de Vasconcelos, L.:* Arch. Microbiol. *99*, 221 (1974).
197. *Nagatani, H.H., Brill, W.J.:* Biochim. Biophys. Acta *362*, 160 (1974).
198. *Benemann, J.R., McKenna, C.E., Lie, R.F., Traylor, T.G., Kamen, M.D.:* Biochim. Biophys. Acta *264*, 25 (1972).
199. *Paneque, A., Vega, J.M^a, Cárdenas, J., Herrera, J., Aparicio, P.J., Losada, M.:* Plant Cell Physiol. *13*, 175 (1972).
200. *Johnson, J.L., Cohen, H.J., Rajagopalan, K.V.:* J. Biol. Chem. *249*, 5046 (1974).
201. *Johnson, J.L., Waud, W.R., Cohen, H.J., Rajagopalan, K.V.:* J. Biol. Chem. *249*, 5056 (1974).
202. *Paschinger, H.:* Arch. Microbiol. *101*, 379 (1974).
203. *Brill, W.J., Steiner, A.L., Shah, V.K.:* J. Bacteriol. *118*, 986 (1974).
204. *Nason, A., Lee, K.Y., Pan, S.S., Ketchum, P.A., Lamberti, A., DeVries, J.:* Proc. Nat. Acad. Sci. USA *68*, 3242 (1971).
205. *Nason, A., Antoine, A.D., Ketchum, P.A., Frazier, W.A. III, Lee, D.K.:* Proc. Nat. Acad. Sci. USA *65*, 137 (1970).
206. *Ketchum, P.A., Cambier, H.Y., Frazier, W.A. III, Madansky, C.H., Nason, A.:* Proc. Nat. Acad. Sci. USA *66*, 1016 (1970).
207. *Pateman, J.A., Cove, D.J., Rever, R.M., Roberts, D.B.:* Nature (London) *201*, 58 (1964).
208. *Ketchum, P.A., Swarin, R.S.:* Biochem. Biophys. Res. Commun. *52*, 1450 (1973).
209. *Ketchum, P.A., Sevilla, C.L.:* J. Bacteriol. *116*, 600 (1973).
210. *Lee, K.Y., Pan, S.S., Erickson, R., Nason, A.:* J. Biol. Chem. *249*, 3941 (1974).
211. *Lee, K.Y., Erickson, R., Pan, S.S., Jones, G., May, F., Nason, A.:* J. Biol. Chem. *249*, 3953 (1974).
212. *Zumft, W.G.:* Ber. Dt. Bot. Ges. *87*, 135 (1974).
213. *Nagatani, H.H., Shah, V.K., Brill, W.J.:* J. Bacteriol. *120*, 697 (1974).
214. *Ganelin, V.L., L'vov, N.P., Sergeev, N.S., Shaposhnikoy, G.L., Kretovich, V.L.:* Dokl. Akad. Nauk SSSR *26*, 1236 (1972).
215. *Ward, M.A., Dalton, H., Mortenson, L.E.:* Bacteriol. Proc. 139 (1971).
216. *Shilov, A.E., Likhtenshtein, G.I.:* Isv. Akad. Nauk SSSR, Ser. Biol. 518 (1971).
217. *Volpin, M.E., Shur, V.B.:* Dokl. Akad. Nauk SSSR *156*, 1102 (1964).
218. *Van Tamelen, E.E., Fechter, R.B., Schneller, S.W., Boche, G., Greeley, R.H., Åkermark, B.:* J. Am. Chem. Soc. *91*, 1551 (1969).
219. *Chatt, J., Pearman, A.J., Richards, R.L.:* Nature (London) *253*, 39 (1975).
220. *Manriquez, J.M., Bercaw, J.E.:* J. Am. Chem. Soc. *96*, 6229 (1974).
221. *Parshall, G.W.:* J. Am. Chem. Soc. *89*, 1822 (1967).
222. *Shilov, A.E., Shilova, A.K., Kvashina, E.F., Vorontsova, T.A.:* Chem. Commun. 1590 (1971).
223. *Chatt, J., Heath, G.A., Richards, R.L.:* Chem. Commun. *1*, 1010 (1972).
224. *Jonas, K.:* Angew. Chem. *85*, 1050 (1973).
225. *Bercaw, J.E., Marvich, R.H., Bell, L.G., Brintzinger, H.H.:* J. Am. Chem. Soc. *94*, 1219 (1972).

226. *Bercaw, J.E., Rosenberg, E., Roberts, J.D.:* J. Am. Chem. Soc. *96*, 612 (1974).
227. *Stiefel, E.I.:* Proc. Nat. Acad. Sci. USA *70*, 988 (1973).
228. *Schrauzer, G.N., Doemeny, P.A.:* L. Am. Chem. Soc. *93*, 1608 (1971).
229. *Schrauzer, G.N., Doemeny, P.A., Kiefer, G.W., Frazier, R.H.:* J. Am. Chem. Soc. *94*, 3604 (1972).
230. *Buchanan, B.B.:* Struct. Bonding *1*, 109 (1964).
231. *Lovenberg, W., ed.:* Iron-Sulfur Proteins. New York and London: Academic Press (1973).
232. *Yoch, D.C.:* Arch. Biochem. Biophys. *158*, 633 (1973).
233. *Zubieta, J.A., Mason, R., Postgate, J.R.:* Biochem. J. *133*, 851 (1973).
234. *Johnson, P.W., Canale-Parola, E.:* Arch. Microbiol. *89*, 341 (1973).
235. *LeGall, J., Dragoni, N.:* Biochem. Biophys. Res. Commun. *23*, 145 (1966).
236. *Shanmugam, K.T., Buchanan, B.B., Arnon, D.I.:* Biochim. Biophys. Acta *256*, 477 (1972).
237. *Yoch, D.C., Arnon, D.I.:* J. Bacteriol. *121*, 743 (1975).
238. *Dunham, W.R., Palmer, G., Sands, R.H., Bearden, A.J.:* Biochim. Biophys. Acta *253*, 373 (1971).
239. *Adman, E.T., Sieker, L.C., Jensen, L.H.:* J. Biol. Chem. *248*, 3987 (1973).
240. *Herskovitz, T., Averill, B.A., Holm, R.H., Ibers, J.A., Phillips, W.D., Weiher, J.F.:* Proc. Nat. Acad. Sci. USA *69*, 2437 (1972).
241. *Mayerle, J.J., Frankel, R.B., Holm, R.H., Ibers, J.A., Phillips, W.D., Weiher, J.F.:* Proc. Nat. Acad. Sci. USA *70*, 2429 (1973).
242. *Frankel, R.B., Herskovitz, T., Averill, B.A., Holm, R.H., Krusic, P.J., Phillips, W.D.:* Biochem. Biophys. Res. Commun. *58*, 974 (1974).
243. *Holm, R.H., Phillips, W.D., Averill, B.A., Mayerel, J.J., Herskovitz, T.:* J. Am. Chem. Soc. *96*, 2109 (1974).
244. *Smilie, R.M.:* Plant Physiol. *40*, 1124 (1965).
245. *Knight, E., Jr., Hardy, R.W.F.:* J. Biol. Chem. *241*, 2752 (1966).
246. *Zumft, W.G., Spiller, H.:* Biochem. Biophys. Res. Commun. *45*, 112 (1971).
247. *Tanaka, M., Haniu, M., Yasunobu, K.T., Mayhew, S., Massey, V.:* J. Biol. Chem. *248*, 4354 (1973).
248. *Andersen, R.D., Apgar, P.A., Burnett, R.M., Darling, G.D., Lequesne, M.E., Mayhew, S.G., Ludwig, M.L.:* Proc. Nat. Acad. Sci. USA *69*, 3189 (1972).
249. *Watenpaugh, K.D., Sieker, L.C., Jensen, L.H., LeGall, J., Dubourdieu, M.:* Proc. Nat. Acad. Sci. USA *69*, 3185 (1972).
250. *MacKnight, M.L., Gillard, J.M., Tollin, G.:* Biochemistry *12*, 4200 (1973).
251. *Mayhew, S.G., Foust, G.P., Massey, V.:* J. Biol. Chem. *244*, 803 (1969).
252. *Yoch, D.C., Valentine, R.C.:* Annu. Rev. Microbiol. *26*, 139 (1972).
253. *Mortenson, L.E., Chen, J.S.:* in Microbial Iron Metabolism (Neilands, J.B., ed.). New York: Academic Press (1974).
254. *Gray, L.T., Gest, H.:* Science (Washington, D.C.) *148*, 186 (1965).
255. *Thauer, R.K., Käufer, B., Zähringer, M., Jungermann, K.:* Eur. J. Biochem. *42*, 447 (1974).
256. *Peck, H.D., Jr., San Pietro, A., Gest, H.:* Proc. Nat. Acad. Sci. USA *42*, 13 (1956).
257. *Cota-Robbles, E.H., Marr, A.G., Nison, E.H.:* J. Bacteriol. *75*, 243 (1958).
258. *Haschke, R.H., Campbell, L.L.:* J. Bacteriol. *105*, 249 (1971).
259. *Kleiner, D., Burris, R.H.:* Biochim. Biophys. Acta *212*, 417 (1970).
260. *Shug, A.L., Wilson, P.W., Green, D.E., Mahler, H.R.:* J. Am. Chem. Soc. *76*, 3355 (1954).
261. *Nakos, G., Mortenson, L.E.:* Biochim. Biophys. Acta *227*, 576 (1971).
262. *LeGall, J., Dervartanian, D.V., Spilker, E., Lee, J.P., Peck, H.D., Jr.:* Biochim. Biophys. Acta *234*, 525 (1971).
263. *Kidman, A.D., Yanagihara, R., Asato, R.N.:* Biochim. Biophys. Acta *191*, 170 (1969).
264. *Nakos, G., Mortenson, L.E.:* Biochemistry *10*, 2442 (1971).
265. *Chen, J.S., Mortenson, L.E.:* Biochim. Biophys. Acta *371*, 283 (1974).
266. *Hardy, R.W.F., Knight, E., Jr., McDonald, C.G., D'Eustachio, J.D.:* in Non-Heme Iron Proteins: Role in Energy Conversion (San Pietro, A., ed.) pp. 275. Yellow Springs, Ohio: Antioch Press (1965).

267. *Bulen, W.A., LeComte, J.R.:* Meth. Enzymology *24*, 456 (1972).
268. *Bulen, W.A., LeComte, J.R., Lough, S.:* Biochem. Biophys. Res. Commun. *54*, 1274 (1973).
269. *Ellfolk, N., Sievers, G.:* Acta Chem. Scand. *27*, 3986 (1973).
270. *Richardson, M., Dilworth, M.J., Scawen, M.D.:* FEBS Lett. *51*, 33 (1975).
271. *Vainshtein, B.K., Harutyunyan, E.H., Kuranova, I.P., Borisov, V.V., Sosfenov, N.I., Pavlovsky, A.G., Grebenko, A.I., Konareva, N.V.:* Nature (London) *254*, 163 (1975).
272. *Dilworth, M.J., Kidby, D.K.:* Exp. Cell Res. *49*, 148 (1968).
273. *Bergersen, F.J., Goodchild, D.J.:* Aust. J. Biol. Sci. *26*, 741 (1973).
274. *Appleby, C.A., Bergersen, F.J., Macnicol, P.K., Turner, G.L., Wittenberg, B.A., Wittenberg, J.B.:* Proc. Internat. Symp. N_2 Fixation, Washington State University Press, Pullman, (1975), in press.
275. *Broughton, W.J., Dilworth, M.J.:* Biochem. J. *125*, 1075 (1971).
276. *Cutting, J.A., Schulman, H.M.:* Biochim. Biophys. Acta *261*, 321 (1972).
277. *Bergersen, F.J.:* Annu. Rev. Plant Physiol. *22*, 121 (1971).
278. *Yocum, C.S.:* Science (Washington, D.C.) *146*, 432 (1964).
279. *Wittenberg, J.B., Bergersen, F.J., Appleby, C.A., Turner, G.L.:* J. Biol. Chem. *249*, 4057 (1974).
280. *Benemann, J.R., Valentine, R.C.:* Adv. Microbial Physiol. *8*, 59 (1972).
281. *Streicher, S.L., Valentine, R.C.:* Annu. Rev. Biochem. *42*, 279 (1973).
282. *Buchanan, B.B., Arnon, D.I.:* Adv. Enzymology *33*, 119 (1970).
283. *Jungermann, K., Thauer, R.K., Leimenstoll, G., Decker, K.:* Biochim. Biophys. Acta *305*, 268 (1973).
284. *Riebeling, V., Jungermann, K., Werdan, K., Heldt, H.W., Thauer, R.K.:* Z. Physiol. Chem. *354*, 1234 (1973).
285. *Haaker, H., de Kok, A., Veeger, C.:* Biochim. Biophys. Acta *357*, 344 (1974).
286. *Reed, D.W., Toia, R.E., Jr., Raveed, D.:* Biochem. Biophys. Res. Commun. *58*, 20 (1974).
287. *Winter, H.C., Ober, J.A.:* Plant Cell Physiol. *14*, 769 (1973).
288. *Jungermann, K., Kirchniawy, H., Katz, N., Thauer, R.K.:* FEBS-Lett. *43*, 203 (1974).
289. *Thauer, R.K., Fuchs, G., Jungermann, K.:* J. Bacteriol. *118*, 758 (1974).
290. *Bothe, H., Falkenberg, B., Nolteernsting, U.:* Arch. Microbiol. *96*, 291 (1974).
291. *Codd, G.A., Rowell, P., Stewart, W.D.P.:* Biochem. Biophys. Res. Commun. *61*, 424 (1974).
292. *Bothe, H.:* Ber. Dt. Bot. Ges. *23*, 421 (1970).
293. *Smith, R.V., Noy, R.J., Evans, M.C.W.:* Biochim. Biophys. Acta *253*, 104 (1971).
294. *Lambeth, D.O., Palmer, G.:* J. Biol. Chem. *248*, 6095 (1973).
295. *Mayhew, S.G., Massey, V.:* Biochim. Biophys. Acta *315*, 181 (1973).
296. *Fisher, R.J., Brill, W.J.:* Biochim. Biophys. Acta *184*, 99 (1969).
297. *Sorger, G.J., Trofimenkoff, D.:* Proc. Nat. Acad. Sci. USA *65*, 74 (1970).
298. *Simon, M.A., Brill, W.J.:* J. Bacteriol. *105*, 65 (1971).
299. *Streicher, S., Gurney, E., Valentine, R.C.:* Proc. Nat. Acad. Sci. USA *68*, 1174 (1971).
300. *Streicher, S.L., Gurney, E.G., Valentine, R.C.:* Nature (London) *239*, 495 (1972).
301. *Kondorosi, A., Barabás, I., Sváb, Z., Orosz, L., Sik, T., Hotchkiss, R.D.:* Nature New Biol. *246*, 153 (1973).
302. *Dixon, R.A., Postgate, J.R.:* Nature (London) *234*, 47 (1971).
303. *Dixon, R.A., Postgate, J.R.:* Nature (London) *237*, 102 (1972).
304. *Cannon, F.C., Dixon, R.A., Postgate, J.R., Primrose, S.B.:* J. Gen. Microbiol. *80*, 227 (1974).
305. *Shanmugam, K.T., Loo, A.S., Valentine, R.C.:* Biochim. Biophys. Acta *338*, 545 (1974).
306. *Shanmugam, K.T., Valentine, R.C.:* Science (Washington, D.C.) *187*, 919 (1975).
307. *Cannon, F.C., Dixon, R.A., Postgate, J.R., Primrose, S.B.:* J. Gen. Microbiol. *80*, 241 (1974).
308. *Stewart, W.D.P., Singh, H.N.:* Biochem. Biophys. Res. Commun. *62*, 62 (1975).
309. *Dunican, L.K., Tierney, A.B.:* Biochem. Biophys. Res. Commun. *57*, 62 (1974).
310. *Bishop, P.E., Evans, H.J., Daniel, R.M., Hampton, R.O.:* Biochim. Biophys. Acta *381*, 248 (1975).
311. *Sutton, W.D.:* Biochim. Biophys. Acta *366*, 1 (1974).
312. *Agarwal, A.K., Mehta, S.L.:* Biochem. Biophys. Res. Commun. *60*, 257 (1974).
313. *Shah, V.K., Davis, L.C., Stieghorst, M., Brill, W.J.:* J. Bacteriol. *117*, 917 (1974).
314. *Cárdenas, J., Mortenson, L.E.:* J. Bacteriol. *123*, 978 (1975).

315. *Seto, B., Mortenson, L. E.:* J. Bacteriol. *120*, 822 (1974).
316. *Daesch, G., Mortenson, L. E.:* J. Bacteriol. *110*, 103 (1972).
317. *St. John, R. T., Brill, W. J.:* Biochim. Biophys. Acta *261*, 63 (1972).
318. *Gordon, J. K., Brill, W. J.:* Biochem. Biophys. Res. Commun. *59*, 967 (1974).
319. *Lepo, J. E., Wyss, O.:* Biochem. Biophys. Res. Commun. *60*, 76 (1974).
320. *Streicher, S. L., Shanmugam, K. T., Ausubel, F., Morandi, C., Goldberg, R. B.:* J. Bacteriol. *120*, 815 (1974).
321. *Tubb, R. S.:* Nature (London) *251*, 481 (1974).
322. *Prival, M. J., Brenchley, J. E., Magasanik, B.:* J. Biol. Chem. *248*, 4334 (1973).
323. *Tyler, B., Deleo, A. B., Magasanik, B.:* Proc. Nat. Acad. Sci. USA *71*, 225 (1974).
324. *Gordon, J. K., Brill, W. J.:* Proc. Nat. Acad. Sci. USA *69*, 3501 (1972).
325. *Doy, C. H., Gresshoff, P. M., Rolfe, B. G.:* Proc. Nat. Acad. Sci. USA *70*, 723 (1973).
326. *Ledoux, L., Huart, R., Jacobs, M.:* Eur. J. Biochem. *23*, 96 (1971).
327. *Carlson, P. S., Chaleff, R. S.:* Nature (London) *252*, 393 (1974).
328. *Child, J. J.:* Nature (London) *253*, 350 (1975).
329. *Scowcroft, W. R., Gibson, A. H.:* Nature (London) *253*, 351 (1975).
330. *Shanmugam, K. T., Valentine, R. C.:* Proc. Nat. Acad. Sci. USA *72*, 136 (1975).
331. *Riederer-Henderson, M. A., Wilson, P. W.:* J. Gen. Microbiol. *61*, 27 (1970).
332. *Postgate, J. R.:* J. Gen. Microbiol. *63*, 137 (1970).
333. *Hamilton, I. R., Burris, R. H., Wilson, P. W.:* Proc. Nat. Acad. Sci. USA *52*, 637 (1964).
334. *Grau, F. N., Wilson, P. W.:* J. Bacteriol. *85*, 446 (1962).
335. *Bergersen, F. J.:* Biochim. Biophys. Acta *130*, 304 (1966).
336. *Stewart, W. D. P., Haystead, A., Pearson, H. W.:* Nature (London) *224*, 226 (1969).
337. *Rippka, R., Neilson, A., Kunisawa, R., Cohen-Bazire, G.:* Arch. Mikrobiol. *76*, 341 (1971).
338. *Venkarataman, G. S.:* Indian J. Agr. Sci. *32*, 22 (1962).
339. *Schollhorn, R., Burris, R. H.:* Fed. Proc. *25*, 710 (1966).
340. *Dixon, M.:* Biochim. Biophys. Acta *226*, 241 (1971).
341. *Yates, M. G.:* Eur. J. Biochem. *24*, 347 (1971).
342. *Benemann, J. R., Yoch, D. C., Valentine, R. C., Arnon, D. I.:* Biochim. Biophys. Acta *226*, 205 (1971).
343. *Yoch, D. C.:* J. Bacteriol. *116*, 384 (1973).
344. *Yoch, D. C.:* J. Gen. Microbiol. *83*, 153 (1974).
345. *Wong, P., Evans, H. J., Klucas, R., Russells, S.:* in Biological Nitrogen Fixation in Natural and Agricultural Habitats (Lie, T. A., Mulder, E. G., eds.) Plant Soil, Spec. vol., pp. 525 (1971).
346. *Fisher, R. J., Wilson, P. W.:* Biochem. J. *117*, 1023 (1970).
347. *Koch, B., Wong, P., Russell, S. A., Howard, R., Evans, H. J.:* Biochem. J. *118*, 773 (1970).

The Uptake of Elements by Biological Systems

J. J. R. Fraústo da Silva*

Centro de Estudos de Quimica Estrutural, Instituto Superior Técnico, Lisbon (Portugal).

R. J. P. Williams

Wadham College, Inorganic Chemistry Laboratory, Oxford. (England)

Table of Contents

I.	The Elements of Biology	69
	1. Introduction	69
	2. Major and Minor Elements in Biology	70
	3. Essentiality of the Elements	73
	4. A Note on Essentiality and Toxicity	74
II.	General Features of the Uptake of the Chemical Elements	76
	1. Selectivity and Biological Environment	76
	2. Available Redox States of Elements	77
	3. Soil-Water-Plant-Animal Relationships	79
III.	Uptake of Cations	81
	1. Generalities: Processes of Uptake	81
	2. Surface (Ion-Exchange) Uptake	81
	3. Uptake by External Chelating Agents	84
	3.1. Charge Type	84
	3.2. Ion Sizes	85
	3.3. Liganding Atom	86
	4. Competitive Binding	86
	4.1. Uptake of Magnesium	90
	4.2. Rejection of Calcium	91
	4.3. Uptake of Zinc	91
	4.4. Uptake of Vanadium	93
	5. Effect of Preferential Coordination Geometry	93
	5.1. Zinc and Copper	93
	5.2. Manganese(III) and Iron(III)	94
	6. Uptake of Transition Metals in Preformed Rings	96
	7. Cooperativity Effects of Metal Clusters	96
	8. Kinetic Control of Metal Chelate Formation	98
	9. Summary: Uptake and Retention of Metal Ions	100
	10. The Blood-Stream as an Example	101

* On sabbatical leave at St. Edmund Hall and Inorganic Chemistry Laboratory, Oxford from The New University of Lisbon.

Acknowledgements: We have benefited from considerable discussions with *Dr. H. A. O. Hill.*
Prof. J. Fraústo da Silva acknowledges a supporting partial fellowship from The Royal Society under the European Science Exchange Programme on nomination by the Academia das Ciencias de Lisboa.

I. The Elements of Biology

1. Introduction

The distribution of the chemical elements in biological systems shows some striking peculiarities. An example, perhaps the most remarkable, is the presence of molybdenum. This element is essential for both plant and animal life, yet it is a rare element. Its immediate neighbours in the second and third transition metal series of the Periodic Table, i.e. Nb, Ta, W, (Tc) and Re, are either absent or extremely rare in biology and have no known functional significance. The much more common element of the same group of the Periodic Table as Mo, i.e. Cr, has an ill defined biological importance, but evidence for an essential function such as that observed for molybdenum has not yet been found.

How did the role of molybdenum evolve? How does living matter obtain the essential supply of this element? Finally, how is the element incorporated into its appropriate site in enzymes, in the face of competition from much larger quantities of somewhat similar elements?

Such problems are not confined to the single element molybdenum, but are general to the involvement of the elements of the Periodic Table in life. In an attempt to appreciate the overall involvement of inorganic elements in biology we examine in this article

(a) the modes of biological capture of given elements,
(b) the modes of incorporation of the elements, and
(c) why biology uses some elements and not others.

The inorganic elements with which we are concerned are not just the transition metals, though they hold preeminence in the eyes of many inorganic chemists involved in the study of biological systems for they lend themselves to physical chemical study by spectroscopic methods, but include the closed shell metals, especially of Group IA and IIA, and a large group of non-metals e.g. Si, P, S, Cl, Se, Br, I.

It is clear that within all groups of elements biological systems concentrate certain of them while rejecting others and that some of these processes are coupled to energy. Nevertheless, no matter whether there is such energy coupling or not, site selectivity is demanded in some steps of the uptake and this site selectivity must be treated independently from the energy input. It is required, then, to understand selectivity of initial uptake independent from any kinetic (energetic) controls. In other words, we wish to show that biology observes rules that are conventional in inorganic analysis: avoidance of interference of one element in the binding of a second, taking into account both the relative concentrations and the binding constants of the free ions in the liquid compartment under examination. If there is a compartment into which two ions can diffuse to reach a high concentration but out of which one of the ions (B) is driven by energised processes, then only limited site selectivity towards the other (A) *in the compartment* is necessary to exclude ion (B) from binding. However, another type of selectivity in binding of (B) must operate to drive this ion out from the compartment; this selectivity must be high and must be a membrane process coupled to energy. Thus biological systems must have a wide variety of ligands in different parts of space so as to concentrate and to separate the inorganic elements.

In principle, initial capture of the elements could be a general process using only a few reagents, but the incorporation of the elements in their proper sites of biological action is required to be very specific. However, it appears from analytical data that cell systems have been highly selective even in the initial uptake process and we shall have the opportunity to comment on some examples given in the present review.

2. Major and Minor Elements in Biology

In principle there is no difficulty in understanding how biological systems get, C, H, O, N, for they are available from the air, but a question of selectivity in their uptake and rejection does arise. How, for instance, do biological systems selectively absorb N_2 rather than O_2 at a metal centre? This is a problem which must be faced (see section IV) and it is worth noting that no general mechanism of selection favouring N_2 is known.

Table 1. Accumulation of Elements by Plants or Animals

Element	Plant	Animal
Al	Club moss, Hydrangea, Tea Plant	
As	Brown Algae	Coelenterates
B	Brown Algae, Plubaginaceae	Sponges
Ba	Rhizopods, Brazil nuts	Coelenterates, Molluscs
Br	Brown Algae	Sponges
Ca	Algae, diatoms	Numerous Marine animal
Cl	Numerous marine and salt desert plants	Soft Coelenterates
Cu	Caryophyllaceae	Annelids, Molluscs, etc.
F	Dichapetalum Cymosum	Vertebrates
Fe	Bacteria	Numerous
I	Brown Algae, Diatoms	Marine Annelids
Li	Thalictrum sp., Cirsium sp.	
Mn	Ferns, Digitalis purpurea	Marine Crustacea
Mo	Papilonacea	
Se	Micrococus Se., Cruciferae, Astragalus Racemosus	
Si	Horse tails	Protozoa, Sponges
Sr	Brown Algae	Alcyonaria, Octocorallia
V	Some fungi	Ascidians
Zn	Thlaspi Calaminare	Coelenterates

Source: General references (1) and (2).

The same can be said for the common biological metals Na, K, Mg and Ca, which are quite abundant and have relatively soluble salts. Reciprocally, there is no difficulty in understanding why Li, Be, Rb, Cs, Sr and Ba are rather uncommon in biology, for although they are elements which have soluble salts and may be easily absorbed, they are of low abundance.

Indeed, since Na and K and, to a lesser degree, Mg and Ca, fulfill such roles as the control of osmotic pressure and the construction of frames which require large amounts of these elements, it is hardly conceivable that rare elements would be chosen for these purposes of relatively low chemical specificity when others are more widely available. In fact, Na, K, Mg and Ca fulfill a number of trivial additional roles, such as charge neutralisation, and although it may still be true that one or all of the rare elements of Groups IA and IIA have such a function, even on a relatively small scale, this has not been detected as yet. (Table 1 shows that there are systems known which do concentrate lithium, barium and so on, but the function of the metals is not known).

Turning to Groups III–V of the Periodic Table, see Fig. 1, we note that no element of the Groups but phosphorus and, perhaps, silicon, in a minor way, is thought to be *generally required* even as a trace element, although boron is recognised as being essential for plants and it is a fact that chromium and vanadium deficiency impairs some physiological functions in laboratory animals. Thus, of the metallic elements of Groups III, IV, V and VI (except Mo), none is required in any known general function by *all* living systems. The fact that elements such as Al and Ti are present in such small amounts does indeed suggest that they are actually rejected by life

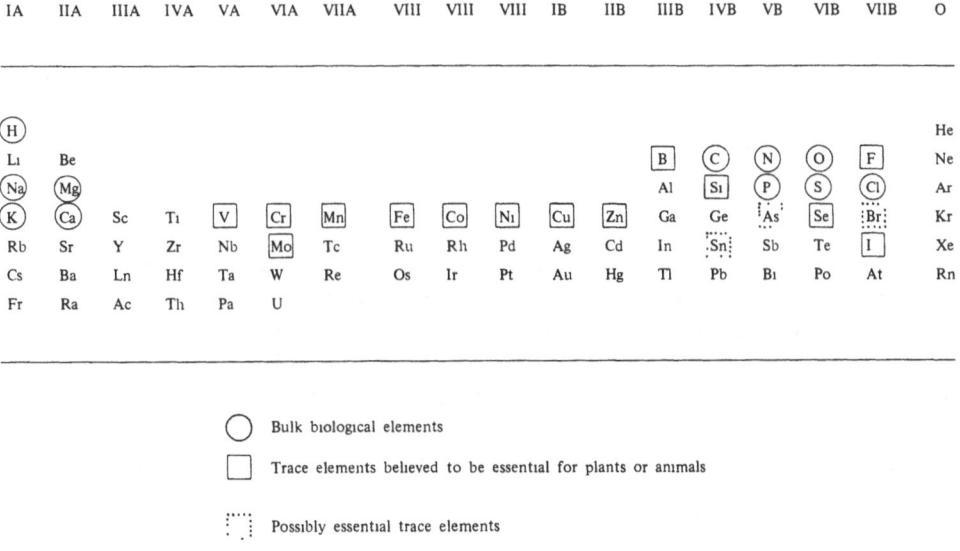

Fig. 1. The Distribution of the Elements Essential for Life in the Periodic Table.

Al	(Si)	P	S	Cl
Sc	Ti	(V)	(Cr)	Mn
Y	Zr	Nb	Mo	Tc
Ln	Hf	Ta	W	Re

Note again that some of these elements are accumulated by specialised plants or animals, Table I, but no clearly defined role for them has yet been discerned.

Leaving the rest of the transition metals aside and looking at the B-subgroups, we find that, while in Group II zinc is certainly essential, there is again a large block of apparently non-essential elements, with the possible exceptions of arsenic and tin.

Zn	Ga	Ge	(As)	Se	(Br)
Cd	In	(Sn)	Sb	Te	I
Hg	Tl	Pb	Bi		

Indeed, not only are these elements apparently non-essential but, if they are present in soluble form above a certain low level, they may be extremely toxic.

Finally, we note that of the elements of the centre of the Periodic Table — Fig. 1 — all the elements from Mn to Cu are used by biological systems, although the roles of Mn, Fe and Cu are overwhelmingly more important than those of Co and Ni. The heavier metals have no known or suggested function

(Cr)	Mn	Fe	Co	(Ni)	Cu
Mo	Tc	Ru	Rh	Pd	Ag
W	Re	Os	Ir	Pt	Au

The immediate reaction of a chemist upon seeing the above figures is to say that it must be the insolubility of the hydroxides (Groups IIIA to VIA) and/or the sulphides (Groups IIB to VIB and VIII) of the elements, together with the low natural abundance of the heavier of them, that makes these elements of low availability, for they must be very difficult to extract. It is partly their highly charged ionic states and partly their high electronegativity which causes their insolubility.

This is equivalent to saying that unless an element is fairly available either as a cation or as an anion at pH = 7, biology does not use it extensively, and if then it is artificially made available to biological systems it will act as a drug or a poison (see note below); thus I, As, Sb, Au, Pt, Pb, Tl, Hg, Cd, Li and Sr have all been used in medicine.

Again, provided no *positive* steps can be seen to have been taken to obtain specific elements, it is likely that, in a competition between cations for a given ligand, it will be an abundant cation which will win overwhelmingly.

Thus, the biology of inorganic elements is largely a biochemistry of the light elements, atomic number less than 30, which form ionic species with low charges, two or less, and have relatively soluble salts. This restricts biology to some 20–25 elements and we would suggest (see below) that if biological evolution has not generated a role for the other elements, this is likely to be due to the fact that any chemical role which they could perform can be performed as well or better by another more readily available element. One may indeed speak of an *economical use of chemical resources* in evolution, the economy being reflected in the energy required for uptake.

There is a final possibility – that an element might have such a chemistry that its presence would be deleterious to the form of chemistry that is life. This would mean that the element would have to be excluded or perhaps accepted in only a very limited way and we would observe this situation by noting that the chemistry which we know to be open to the element does not appear in biology. A common element for which this situation appears to have arisen is chlorine for it is accepted as chloride but it is almost never made into combined chlorine in organic chemicals although man has shown that very interesting compounds can be made by chlorination of organic molecules. This situation invites the question 'Are organo-chlorine compounds and biology incompatible?' and if so, 'Do we not need to look at the chemical industry with questioning eyes?'

3. Essentiality of the Elements

When we asked above why so few elements are *essential* even at the trace level, and supposing biology could have been ingenious enough to get all elements, the first answer was to refer to availability. A second point was the overall redundancy in the chemistry of the elements. The number of elements which exist in any quantity on the surface of the earth, is a consequence of *nuclear* stability and does not reflect chemical (electronic) diversity. A Periodic Table, Fig. 1, is a Table of redundancies by its very nature, thus we can expect that one or two elements from each Periodic Group will be able to carry out chemistry encompassing that of all other members. For example K, Na, Ca, Mg, are required and Rb, Cs, Ba, Sr are redundant. All lanthanides are redundant, since much, if not all, of their chemistry can be done by calcium. Iron can do most transition metal chemistry, zinc most acid/base chemistry. If we combine redundancy in chemistry with the difficulties of abundance and extraction of many elements, i.e. availability, it is reasonable to suppose that only a few elements are indeed essential. Thus we reach the conclusion, similar to that in discussions of many other features of life, that its *observed* features (here essentiality of the elements) is a consequence of *economical functionality*, the hall-mark of existence in a competition, usually referred to as "the survival of the fittest".

We note however that the "survival of the fittest" requires an avoidance of risk. Is it for this reason that the metal ion which is used in acid catalysis in enzymes is zinc and not copper? The Cu(II) ion is undoubtedly a better acid but it also catalyses redox reactions and the dual character of Cu(II) could generate risk in the reactions catalysed. The economy here is in the unambiguous simplicity of each system for each type of function.

4. A Note on Essentiality and Toxicity

The two extreme patterns of the effect of different concentrations of a given chemical upon life processes is shown in Fig. 2. There is an initial region where the chemical is either present in too low concentration for full activity (i) or in too low concentration to be harmful (ii). This region is followed in case (i) by an optimum dosage, but, at higher levels, *all* chemicals are deleterious in their effects — Table 2. Excessive uptake is as deleterious as deficiency. Elements which show effects following curve (i) are usually found to be essential while elements following curve (ii) are likely to be poisons. Careful use of the poisons can yield drugs, for elements which show this behaviour can compete with sites for the other elements or may bind a quite novel set of ligands. In this case, the corresponding diagram will be intermediate between (i) and (ii). It is worth noting that essential biological elements are all in Class (a) of Chatt's (chemical) classification, while most seriously harmful elements fall in Class (b). The reasons for this have been discussed (*1*).

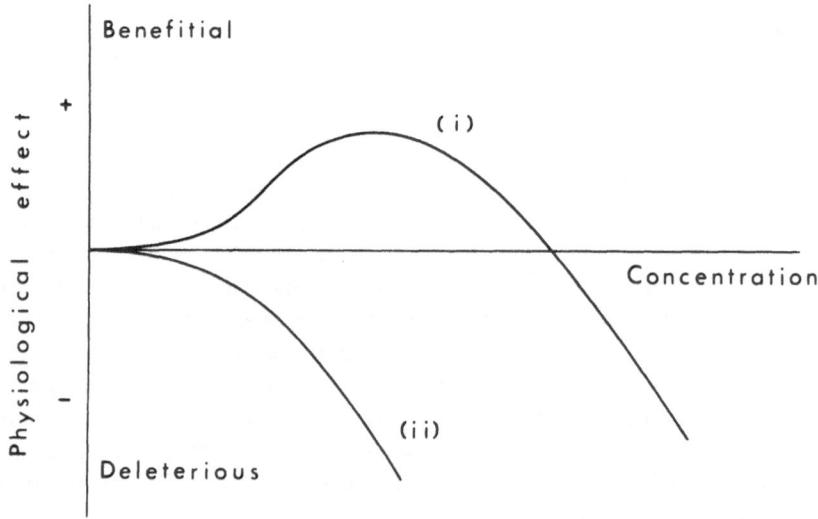

Fig. 2. Physiological Effects of Essential Elements and of Poisons according to dosage.

Table 2. Some Diseases Associated with particular Elements

Element	Disease	
	Deficiency	Excess
Cd	—	Nephritis
Co (B_{12})	Anaemia	Polycythemia
Cr	Abnormal Glucose Metabolism	—
Cu	"Swayback"	Wilson's disease
F	Dental caries	Fluorosis
Fe	Anaemia	Hemochromatosis
I	Goitre	—
Mn	Skeletal Abnormalities	Ataxia
Pb	—	Anaemia
Se	Liver necrosis	"Blind staggers"
Si	Skeletal Abnormalities	Silicosis
Zn	Dwarfism	Metal-fume fever

Adapted, in part, from *Ulmer, D. D.:* Fed. Proc. *32*, 1758, (1973)

II. General Features of Uptake of the Chemical Elements

1. Selectivity and Biological Environment

The uptake of an element is grossly affected by the nature of the compartment in which is is initially and by the compartment into which it is to be extracted for there is no absolutely fixed biological environment for a metal ion. The pH, the ionic strength, the effective redox potential, quite apart from the solvent and chemical composition of biological media, vary quite considerably. We may suppose generally that inside a cell the pH is 7 ± 1, the ionic strength is 0.3 ± 0.1 M, the redox potential is < 0.0 and could be as low as − 0.4 volts, and that the solvent is usually water. Outside the cell, in the blood stream of an animal and probably in such water as sea-water, the pH is 7 ± 1, the ionic strength is 0.3 ± 0.1 M but the redox potential may be as high as + 0.5 to 0.8 volts, and the solvent is water. Inside special body fluids, such as gastric juices, the pH may be as low as approx. 1.0; in salt glands or in the vesicles of adrenals ionic strength could be as high as ≫ 1.0 M. In plant tissues all these variables change, from vacuole to tonoplast (cell) fluids, to chloroplast fluids, to mitochondrial fluids. It is not possible to describe the selective uptake of ions into such different media in detail here, but general principles can be discerned for the usual cases of pH = 7 and ionic strength ∼ 0.3 M.

The pH = 7 has a severe influence upon the uptake reactions of cations or anions. There are two main effects: (1) that of the degree of dissociation of a potential ligand, e.g.

$$R-OH \rightleftarrows RO^- + H^+$$

$$RNH_3^+ \rightleftarrows RNH_2 + H^+$$

where anions RO^- and molecules RNH_2 could be uptake centres for cations and the species $R-OH$ and RNH_3^+ could be uptake centres for anions. The pK_a values of these groups are equal in importance with (2) the hydrolytic reactions of the species to be absorbed, e.g.:

$$Fe^{3+} + 6 H_2O \rightleftarrows Fe(OH)_3 + 3 H_3O^+$$

$$H_2PO_4^- + H_2O \rightleftarrows HPO_4^{2-} + H_3O^+$$

It is clear that for uptake the proton is a competing and controlling cation and that the hydroxide ion is a competing and controlling anion. At pH = 7 the ligands with $pK_a \gg 7$ will be better for anion uptake while the ligands with $pK \ll 7$ are favoured for cation uptake. If RNH_2 or RS^- ligands ($pK_a > 7$) are to capture metal ions they must have much greater metal binding constants relative to $R-CO_2^-$, $ROPO_3^{2-}$ and imidazole ($pK_a \ll 7$). Of course all reactions are also controlled by the redox states of the elements involved.

2. Available Redox States of Elements

There is a curious difference between transition metals of the first transition series and non-metals with respect to their oxidation state diagrams — Fig. 3 and 4. If we limit the range of potentials to the region between the H_2/H^+ and the O_2/OH^- potentials, then a transition metal has a short range of stable oxidation states in water, e.g. iron

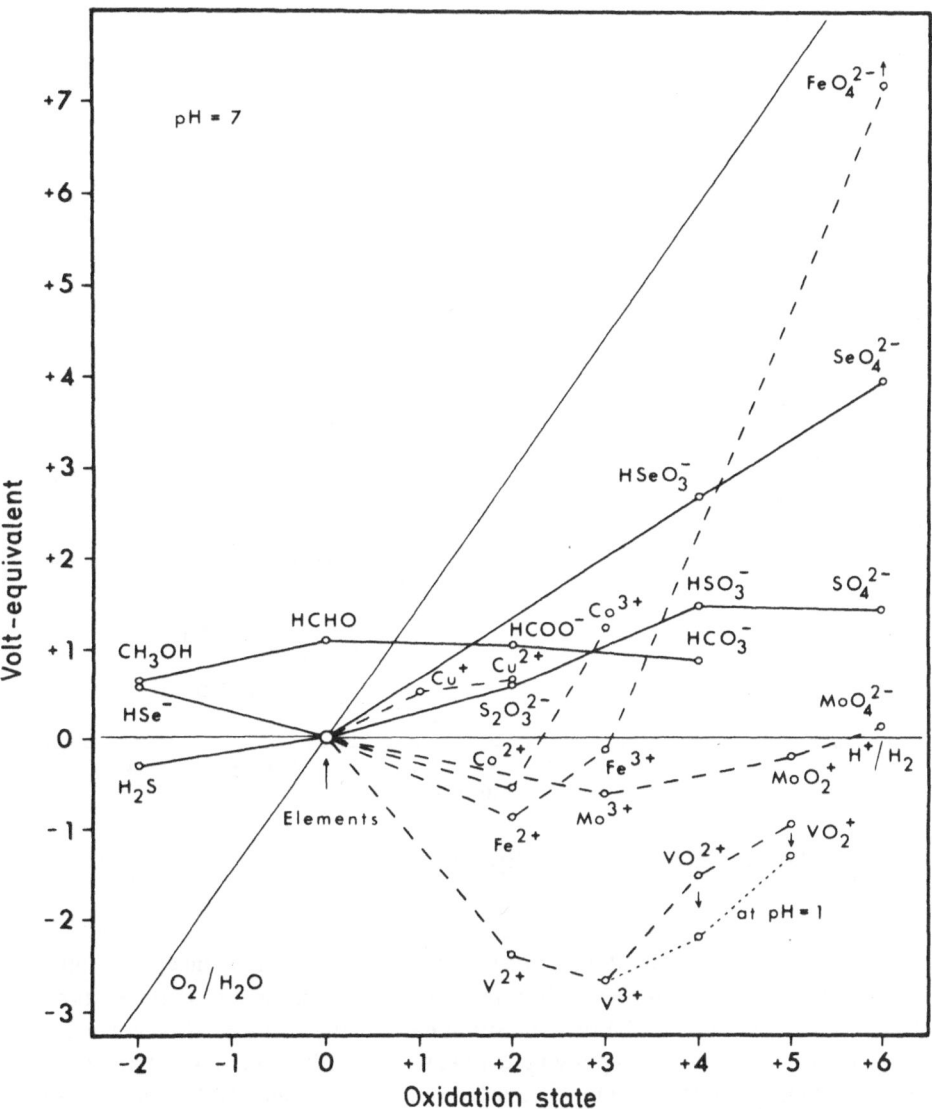

Fig. 3. Oxidation State Diagram for some metals and non-metals at $pH = 7$.

77

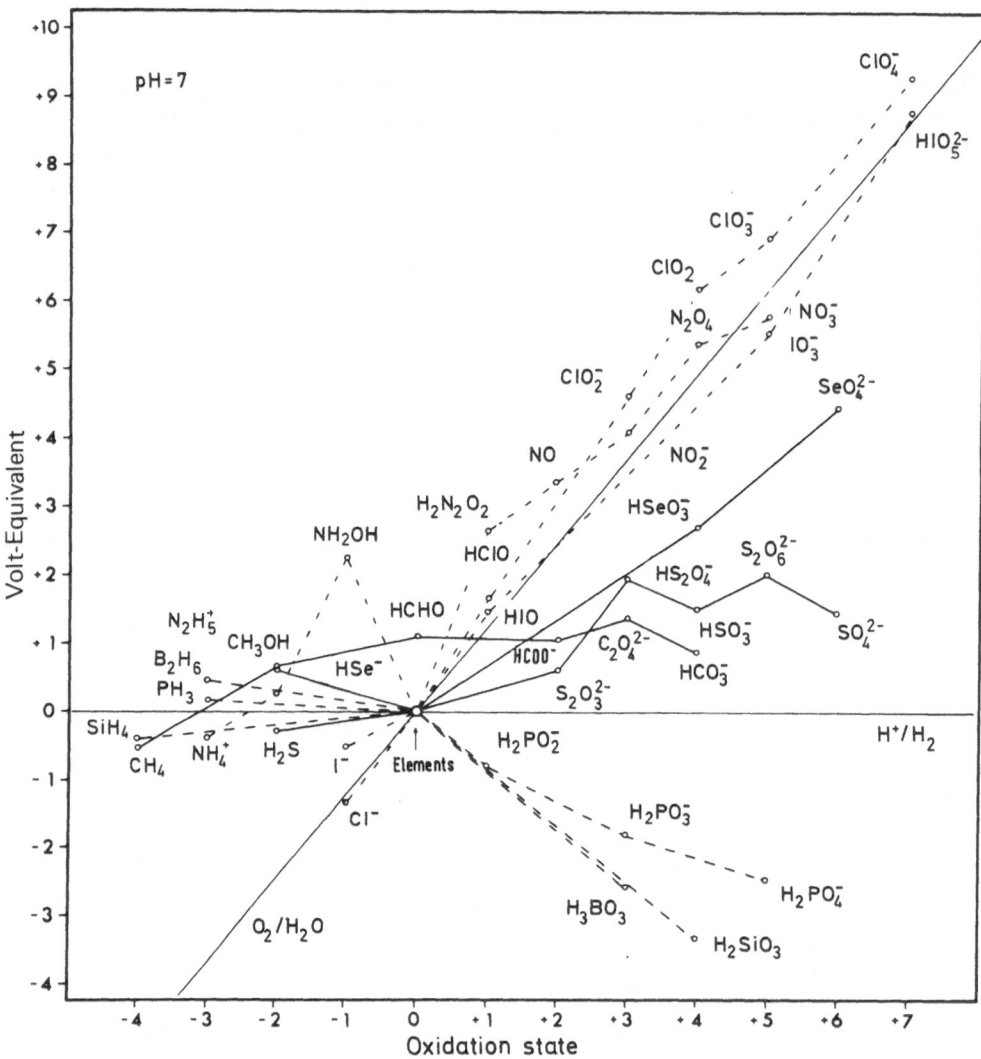

Fig. 4. Oxidation State Diagram for non-metals at *pH* = 7.

(2 and 3) copper (1 and 2), but a non-metal may have a very wide range, e.g., sulphur (− 2 to + 6), carbon (− 4 to + 4). This difference arises from the different electrons involved in the oxidations. In the first transition metal series is is the *3d* core which is ionised and this process is of rapidly increasing energy with increasing charge and cannot be readily compensated by chemical binding. In non-metals the ionisations are of *sp* electrons and here compensation by chemical bond formation is possible. This difference will be seen later to allow a very different chemistry of uptake. The only

element of biological interest for which these generalisations fail is *molybdenum*, which has a significance all of its own, since it is a second row transition metal element and, as such, behaves more like a non-metal in that it shows a wider variety of oxidation states within the restricted range of potentials. The case of vanadium is also rather interesting and affords a special example which nature can use to provide unexpected solutions for different chemical problems; we will deal with this example later.

From the fact that the redox potential in biology must lie between approximately $+0.8$ to -0.5 volts there are non-metal elements with one very stable state: Si/O, P/O, B/O and Cl^-, Br^- etc., at the extremes of their available redox states. Other elements have very unstable states at the extremes, for example N/O, Cl/O, I/O, P/H, (Se/H), etc. Thus biology uses only a relatively small number of redox switches within the available elements, such as $CO_2 \rightarrow CH_4$, $SO_4^{2-} \rightarrow SH_2$, $N_2 \rightarrow NH_3$ (note that NO_x states are always run down to HN compounds) and $SeO_4^{2-} \rightarrow Se$ (note that SeH compounds are not very probable.) This division between elements which can readily undergo *reversible* redox switches (C, S, Se,) and those which are restricted to low (N) or high (P, Si, B) oxidation states is helpful when considering modes of separation.

It is also important to observe the oxidation conditions under which an element can be absorbed and then subsequently converted to another oxidation state for final capture.

3. Soil-Water-Plant-Animal Relationships

If we are to appreciate the skill employed by living systems in the uptake of the elements we must study first the forms in which they are available in the natural surround of life, taking into account all the above chemical limitations imposed by the major medium, water, the presence of oxygen, and the restrictions of the small temperature and pressure ranges of the solvent water upon the possible chemical steps by which life can handle the elements. A safe starting premise is the availability of the elements in soil water or the sea, where the required elements are then in the forms shown in Tables 3 and 4. These forms are also likely to be those in the blood stream of an animal and the sap of a plant. In the following sections these media will all be treated together.

Some of the peculiarities of the redox potentials of the elements appear in these data. Thus, in the transition series chromium is only present as a minute trace in water,

Table 3. Forms in which main biological elements occur in aerated soil water, the sea or in blood plasma (simple species).

Cations	Anions	Neutral
H_3O^+, NH_4^+	HCO_3^-, CO_3^{2-}, NO_3^-	$B(OH)_3$
Na^+, K^+	$PO_4H_2^-$, PO_4H^{2-}	CO_2, $SiO_2 \cdot (H_2O)_n$
Mg^{2+}, Ca^{2+}	OH^-, F^-, Cl^-, Br^-, I^-, SO_4^{2-}	N_2, NH_3, O_2

79

Table 4. Inorganic speciation of some trace metals in marine systems at 25 °C, 1 atm pressure, 35% 0 salinity, and $pH = 8$

Metal	Major Forms
Al^{3+}	$Al(OH)_3^0$
Cr^{3+}	$Cr(OH)_2^+$ 85%, CrO_2^- 14%
Mn^{2+}	Mn^{2+} 58%, $MnCl^+$ 30%, $MnSO_4^0$ 7%
Fe^{2+}	$FeOH^+$ 84%, $FeCl^+$ 7%, Fe^{2+} 7%
Fe^{3+}	$Fe(OH)_3^0$ 95%, $Fe(OH)_2^+$ 5%
Co^{2+}	Co^{2+} 54%, $CoCl^+$ 31%, $CoCO_3^0$ 7%, $CoSO_4^0$ 7%
Ni^{2+}	Ni^{2+} 53%, $NiCl^+$ 31%, $NiCO_3^0$ 9%, $NiSO_4^0$ 6%
Cu^{2+}	$CuOHCl^0$ 65%, $CuCO_3^0$ 22%, $CuCl^+$ 6%, $CuOH^+$ 4%
Zn^{2+}	$ZnCl^+$ 44%, Zn^{2+} 16%, $ZnCl_2^0$ 15%, $ZnOHCl$ 13%
Mo(VI)	MoO_4^{2-}
Ag^+	$AgCl_4^{3-}$ 54%, $AgCl_3^{2-}$ 24%, $AgCl_2^-$ 17%
Cd^{2+}	$CdCl_2^0$ 50%, $CdCl^+$ 40%, $CdCl_3^-$ 6%
Hg^{2+}	$HgCl_4^{2-}$ 66%, $HgCl_3Br^{2-}$ 12%, $HglCl_3^-$ 12%
Pb^{2+}	$PbCl_2^0$ 42%, $PbCl^+$ 19%, $PbOH^+$ 10%, $PbOHCl^0$ 9%

Table extracted from The Nature of Sea Water, p. 22 (ed. E. D. Goldberg). report of the Dahlem workshop, Berlin 1975.

as the redox potentials restrict chromium to Cr(III), so that it cannot be obtained as a soluble cation, e.g. Cr^{2+}, or as a soluble anion, such as CrO_4^{2-}. The concentration of iron, mainly as Fe^{2+}, is much higher, as there is a considerable amount of organic material present which is capable of reducing iron to this oxidation state.

The above environmental availability restricts biology to some extent, but there is still the possibility of bringing an element, e.g. iron, into contact with an appropriate set of ligands to optimalise uptake (and subsequently function) while at the same time preventing competition from other cations for the same site, e.g. competition for iron sites by copper ions. Sufficient is now known to make suggestions about such selectivity based upon restrictions of both a thermodynamic and a kinetic kind.

III. Uptake of Cations

1. Generalities: Processes of Uptake

The steps of uptake of either an anion or a cation can be generalised. The simplest step is surface absorption which we write for cations, M

$$M + \text{surface} \rightleftharpoons M\text{ surface} \tag{a}$$

An alternative initial step is for the biological system to provide an *external* chelating agent, L, in order to capture the element

$$M + L \rightleftharpoons ML \text{ (external solution)} \tag{b}$$

Both steps must be followed by a passage of the captured element deeper into a cell

$$M \text{ surface} + L^1 \rightleftharpoons ML^1 + \text{surface}$$

$$ML \qquad + L^1 \rightleftharpoons ML^1 + L$$

There can be changes of redox state prior to the transfer of the element into the cell or subsequently. Finally the chelate ML^1 may be but a way of handing M on to some final site for M

$$L^{11} + ML^1 \rightleftharpoons ML^{11} + L^1$$

Not all of these reactions will come to rapid equilibrium and we shall have to deal with both thermodynamic and kinetic controls of both acid/base and redox equilibria in the overall uptake.

2. Surface (Ion Exchange) Uptake

The simplest mode of uptake of ions is by absorption on the exposed plant surfaces e.g. cell walls of bacteria, root hairs, or even the leaf of a plant and the skin of an animal. The study of this absorption has become an important aspect of ecology, and in particular lichens are used as test plants for the study of atmospheric pollution by such elements as nickel, lead and strontium (2). From the blood stream of an animal or the sap of a plant a similar uptake is possible onto the face of the cells lining those capillaries which carry the blood and the sap.

As the uptake is onto the surface of the specimen where binding is weak and exchange is rapid the process has been treated as an equilibrium ion-exchange process;

81

Table 5. Affinity of metal ions for some plant tissues

Ion	Species	Organ or tissue	$\log(1/_{Km})$	$\log K$
K^+	Hordeum vulgare (Barley)	Excised roots	1.68	–
K^+	Zea mays (Maize)	Leaf tissue	1.42	–
K^+	Elodea densa (Waterweed)	Leaf (whole)	1.00	–
Rb^+	Hordeum vulgare (Barley)	Excised roots	1.77	–
Rb^+	Zea mays (Maize)	Leaf tissue	1.82	–
Sr^{2+}	C. alpestris (Lichen)	Powdered plant	–	1.95
Ni^{2+}	U. muhlenbergii	Powdered plant (pKa = 5.40)	–	3.30
Ni^{2+}	U. muhlenbergii	Powdered plant (pKa = 2.85)	–	3.05
Cu^{2+}	Sugar cane	Leaf	1.87	–
Zn^{2+}	Sugar cane	Leaf	1.95	–
Mn^{2+}	Sugar cane	Leaf	1.79	–

K_m – Michaelis' Constant
K (for acid centres in Lichens) = $(MA^+)/(M^{2+})\cdot(A^-)$
Ref. – For Lichens – *Tuominen, Y.*: Ann. Bot. Fenn. 4, 1, (1964). *Nieboer, E., Puckett, K. J., Grace, B.*: Canad. J. Bot. 54, 724 (1976).
For other plants – adapted from: *Clarkson, D.*: Ion transport and cell structure in plants, pp. 282–283, Maidenhead: MacGraw Hill Book Co. 1974.

by conventional treatment of such processes binding constants for different ions to the surface can be calculated. Using slightly different approaches different authors have arrived at similar constants some of which are given in Table 5. It would appear from the values presented that the binding groups have an acid dissociation constant, pK_a, of from 2 – 6 indicating carboxylate or possibly phosphate ester sites. Binding constants of metals to these centres do not show a wide variation (Table 6) as would be expected for such sites, and their values ($\log K_M$) are generally small. In these circumstances it will be unusual for the surface centres to be blocked even by metal ions which are very abundant, e.g. calcium and magnesium, and therefore there is general uptake based on the available free ion concentration.

Deeper within cell walls there are much stronger binding sites than those on the very surface and several lines of evidence indicate that these metal binding sites undergo slower exchange and may become saturated. Thus it is known that the walls of bacteria such as *E. Coli* are stable in fresh water but not when EDTA is added to the solvent: calcium (and magnesium) is slowly extracted from the sites and the walls of the cell tend to fall apart (3). Again many plant wall saccharide structures are stabilised by calcium which cannot be removed easily. A binding constant internal to the wall of $\log K_M \geqslant 6$ is probable and the function of the metal ion is to cross-link various acidic groups (again carboxylate, sulphate and phosphate centres). The cross-linked polymers generate slow exchange sites. Detailed study has been made of these systems (4).

Consider first the case where such sites are surrounded by the simplest aqueous media e.g. roots in soil or plants in the sea. In the absence of heavy pollution, this type of site selects calcium ions. The reason for this is obvious enough for the selection is based upon

$$\text{Amount Bound } \alpha \text{ [concentration]} \times [\log K_M]$$

Table 6 shows that $\log K_M$ for different cations such as calcium, nickel, and copper does not vary greatly for carboxylate, sulphate and phosphate binding centres so that selection amongst these ions rests upon the concentration term. The concentration of calcium usually overwhelms that of any other divalent ion, except magnesium, by several orders of magnitude; as $\log K_M$ for calcium exceeds that of magnesium at these multi-coordinating centres, calcium is taken up on most cell surfaces. Magnesium therefore is absorbed to a lesser extent as stereochemical constraints make its binding constant smaller than that of calcium. An exception to this argument will be the case of deposits on surfaces from the air, when the preferential uptake is of the most common pollutant, e.g. nickel by lichens around nickel works (5).

The above account of surfaces makes it clear that special mechanisms may be required for the absorption of ions other than calcium. Conversely it is also worth noting that special mechanisms will be required for the rejection of calcium.

Just as the external surfaces of cells do not provide sites of great selectivity for cations, so the anion-binding sites based on positively charged centres of the surfaces are likely to be equally unselective and in sea water, for example, these sites could well be overwhelmingly bound by the predominant anions Cl^- and SO_4^{2-}. There is then a need for different uptake and rejection devices so as to make capture of the rarer elements possible. Here there are clearly two possibilities: (a) The cell can send out scavenger chemicals to bind a particular element and pass the combined form back through the cell membrane, or (b) the cell can make a suitable scavenger-compound bound in the cell membrane surface which can transport ions. Examples of both situations are known, e.g. small chelating agents thrown out to catch iron (6) and membrane proteins designed to transport sulphate (7).

Table 6. Stability constants ($\log K_M$) of alkaline-earth and first series transition metal ions with some carboxylate, phosphate and sulphate ligands (T = 20 or 25 °C).

Ligand	$\log K_M$									
	Mg^{2+}	Ca^{2+}	Sr^{2+}	Ba^{2+}	Mn^{2+}	Fe^{2+}	Co^{2+}	Ni^{2+}	Cu^{2+}	Zn^{2+}
Acetate	1.25	1.24	1.19	1.15	1.40	–	1.46	1.43	2.23	1.57
Malonate	1.95	1.85	(1.25)	1.34	(2.3)	(2.5)	2.98	3.30	5.55	2.97
Phosphate	1.60	1.33	1.0	–	2.58	–	2.18	2.08	3.2	2.4
Pyrophosphate	5.4	4.9	(4.7)	(4.6)	–	–	6.1	6.98	7.3	(5.1)
Methylphosphate	1.52	1.49	–	–	2.19	–	2.00	1.91	–	2.16
Sulphate	2.20	2.31	2.1	2.3	2.0	2.3	2.47	2.40	2.4	2.3
Thiosulphate	1.84	1.98	2.04	2.21	1.95	2.17	2.05	2.06	–	2.30

Source: Stability constants. The Chemical Society Special Publications No. 17 and 25. London, 1964, 1971.

3. Uptake by External Chelating Agents

The uptake by an agent thrown out by a cell or by the outer surface of a membrane must take place at around $pH = 7$ and at a redox potential around that of oxygen, i.e. + 0.8 volts. In such media the selection of metal ion from metal ion (or anion from anion) can be based on the pH and the charge type to which the metal is driven by oxidation. The type of ligand which can effect capture is however limited by the same redox potential and pH. Hence, we must inspect the possible available cations (see section II. 2. above) and the possible available ligand donors which can bind them, taking due note of E^0 and pK_a values. For example, we can neglect RS^- ligands as external scavengers for they will be oxidised to RS–SR or other sulphur species which do not bind metals well. In turn the ligands can select cations by virtue of charge, size, covalence, etc.

3.1. Selection by Charge-Type

In water under the conditions described above there is a distinction between the cations of higher and lower oxidation state which follows the divisions of the Periodic Table. From Sc to Zn, the element iron and all elements before it will normally be in oxidation state $\geqslant 3 +$ (see above the comments on the occurrence of Fe^{2+} in natural waters); cobalt and all ions after it to zinc will be in oxidation state $2 +$. Thus, we need to ask which ligands will select $3 +$ ions in preference to $2 +$ ions, given the above limitation on the ligands.

The obvious answer is small anions and we know from analysis that this means particularly small first row anions: $> N^-$, $R-O^-$, F^-. Cl^- binds too weakly, RS^- is absent, R_3C^- is absent and so on. The pK_a value of $> NH$ is very high, so that at $pH \sim 7$ there is little $> N^-$; F^- (and Cl^-) cannot be combined in an organic group or other unit and are therefore unlikely to be involved in capture devices. This leaves RO^-, which can occur in enolates, phenolates, etc. and these are the anions which in fact selectively remove trivalent cations. We then appreciate why the uptake and binding of iron can be as trivalent RO^- complexes.[1] Examples appear in Table 7.

It might be asked what of ligands such as OH^- and O^{2-}, the hydrolytic ligands which are in large part the precipitating anions for trivalent ions in water. The answer is that they fail to compete effectively with enolate and phenolate in water at $pH = 7$ as these organic anions have much lower pK values, of around 10.0, but see also section III. 7. Chromium should also appear as Cr(III) and it may be that it is a

---

[1] It is worth noting that the binding of Fe^{3+} preferably to Fe^{2+} lowers the redox potential of the Fe^{3+}/Fe^{2+} couple facilitating oxidation to the Fe(III) state. In the absence of such a reagent for the uptake of Fe^{3+} then uptake of Fe^{2+} is the more probable mode of absorption of iron (as simple Fe^{3+} ions are precipitated as hydroxide) and there are some scavenger chemicals for Fe^{2+}. However these are only selective for Fe^{2+} against other metal ions if binding of Fe^{2+} involves a spin-state change, see later.

Table 7. Sample R–OH Ligands which bind Fe^{3+}

Catechol	Acetylacetone	Phenol
Chromotropic Acid	Citric Acid	Oxine
Pyrogallol	Lactic Acid	Conalbumin
Salicylic Acid	Desferrioxamines	Transferrin
Tiron	Desferrichromes	Superoxide Dismutase
Kojic Acid	Humic Acids	Hemerythrin

deficiency of analysis that has failed to find Cr(III) chelates especially phenolates in biology.[2]) However, relative to the stability of the oxides, Cr^{3+} complexes, even of phenolates, are somewhat less stable than Mn^{3+} and Fe^{3+} complexes. Moreover they cannot enter solution leaving the soil as the M^{2+} state, for the Cr^{2+} ion is too reducing. A further barrier to Cr^{3+} uptake is its stability, both thermodynamic and kinetic, in octahedral holes. Thus many chemical factors militate against the uptake route available to both Mn and Fe

$$\text{Soil} \rightarrow M^{2+} \xrightarrow[\text{outside cell}]{\text{biology}} M^{3+} \text{(carrier)} \xrightarrow[\text{cell}]{\text{in}} M^{2+}$$

Further back in the first row of transition metal ions the problem is aggravated, for Ti and Sc are even less able to follow this pathway. Note, however, that the hydrolysis of vanadium to the oxocation and of molybdenum to the anion allow the following possible paths, provided that pH of the medium is suitable (see later):

$$\text{Soil} \rightarrow VO^{2+} \rightarrow \text{carrier} \xrightarrow[\text{cell}]{\text{in}} V(III)$$

$$\text{Soil} \rightarrow MoO_4^{2-} \rightarrow \text{carrier} \xrightarrow[\text{cell}]{\text{in}} Mo(V) \text{ etc.}$$

Molybdenum and vanadium uptake are discussed again later.

3.2. Ion Sizes

The sizes of atomic ions vary grossly from the smallest 0.05 nm to the largest 0.15 nm radius. It was proposed many years ago that ring chelates could be designed to select for metal ions on the basis of a compatibility between the cavity size of the ring chelate and the ion size. This expectation has been fulfilled in the last ten years with the result that there are now very many classes of ligand of this type; the selectivity which

[2]) Note that the glucose tolerance factor isolated from brewer's yeast appears to contain a chromium complex of nicotinic acid, cysteine, glutamic acid and glycine (8)

Table 8. Metal binding constants ($\log K_m$) of macrobicyclic ligands in aqueous solution ($T = 25\,^\circ$C)

Ligand

		Cavity size	Li⁺	Na⁺	K⁺	Rb⁺	Cs⁺	Mg²⁺	Ca²⁺	Sr²⁺	Ba²⁺
		Å	(0.60)	(0.95)	(1.33)	(1.48)	(1.69)	(0.65)	(0.99)	(1.13)	(1.35)
	m	n									
I	0	1	0.8	4.30	2.80	< 2	< 2	< 2	–	2.80	< 2	< 2
II	1	0	1.15	2.50	5.40	3.95	2.55	< 2	< 2	6.95	7.35	6.30
III	1	1	1.4	< 2	3.90	5.40	4.35	< 2	< 2	4.40	8.00	9.50
IV	1	2	1.8	< 2	< 2	2.2	2.05	2.2	< 2	~ 2	3.40	6.00
V	2	1	2.1	< 2	< 2	< 2	< 0.7	< 2	< 2	< 2	~ 2	3.65
VI	2	2	2.4	< 2	< 2	< 2	< 0.5	< 2	–	–	–	–

Source: *Lehn, J. M.:* Structure and Bonding, Vol. 15, p. 1. Berlin–Heidelberg–New York: Springer 1973.

can be achieved is shown in Table 8. We return to a discussion of ring chelates later (*Note*: Volume 16 of Structure and Bonding is devoted to this topic).

3.3. Liganding Atom

A totally different basis of selectivity is the use of different donor atoms in the chelating agent i.e. N, O, or S. While A- group metal ions have high affinity for O-ligands, the transition metal ions and B- subgroup metal ions bind equally well or better to N- and S- donors. This topic is well reviewed (*9, 10, 11*).

We must now put together all the cations and all the possible ligands and examine their competition on the basis of charge, size and preference for liganding donor, i.e. covalence.

4. Competitive Binding

Let us examine first the possible mechanisms of selection amongst mono- and divalent cations within a cell as these ion states are often freely available to the cell.

Selective uptake in all cases must be based upon the competitive binding as described by the equation

$$\text{Uptake} \propto \text{Concentration} \cdot K_M$$

As the concentration of free cations falls dramatically

K^+, Na^+	Mg^{2+}, Ca^{2+}	Zn^{2+}	Cu^{2+}
10^{-1} M	10^{-3} M	$< 10^{-7}$ M	$< 10^{-10}$ M

clearly the effective binding constants of ligands at $pH = 7$ must be chosen as in Table 9 where K_M is the binding constant effective at $pH = 7$. The ligands which fall into these classes (if properly designed) are ligand I: ether, alcohol and carbonyl oxygen; ligand II: carboxylate or phosphate $- O^-$ (not phenolate); ligand III: nitrogen or sulphur donors; All of them must have a low pK_a, i.e. not greatly in excess of $pK_a = 7$, so that they are available at physiological pH.

Table 9: Differentiating ligands for cations

Ligands	Stability constant (K_M) and product $K_M[M]$ (uptake factor)	Na^+, K^+	Mg^{2+}, Ca^{2+}	Zn^{2+}, Cu^{2+}	Examples
Ligand I	K_M	$\geqslant 10$	$\leqslant 10^2$	$\leqslant 10^6$	large O–hole of
	Uptake factor	$\geqslant 1$	$\leqslant 0.1$	$\leqslant 0.1$	macrocycle
Ligand II	K_M	0.1	$\geqslant 10^3$	$\leqslant 10^6$	di- or tri-
	Uptake factor	$\leqslant 0.1$	$\geqslant 1.0$	$\leqslant 0.1$	carboxylate
Ligand III	K_M	0.1	$\leqslant 10^2$	$\geqslant 10^6$	Nitrogen ligands
	Uptake factor	$\leqslant 0.1$	$\leqslant 0.1$	$\geqslant 1.0$	

One way in which the separations of Na^+ and K^+, Mg^{2+} and Ca^{2+} are achieved on the basis of charge and size was described above and more details are given below but the separation of the transition metals and zinc together with the evidence of uptake of such metals as Cd^{2+} requires a subtle chemistry for many of these cations are of similar size, charge and covalence. For many of these essential elements there is not only a problem of initial uptake due to very different abundance but also a problem of incorporating the metal ion into the correct functional site. To examine this aspect of uptake for *trace* elements it may well be essential that there is an excess of ligands over some metal ions so that the concentration of each such metal is controlled by a binding site designed for it and from which other similar metal ions are excluded in a doubleheaded competition. The situation can thus be represented as:

$$L_1 + L_2 + M_1 + M_2 \longrightarrow M_1 L_1 + M_2 L_2$$
$$L_1 + M_2 \longrightarrow \text{excluded by lack of } M_2$$
$$L_2 + M_2 \longrightarrow \text{excluded by lack of } M_1$$

87

This is equivalent to saying that the 'conditional stability constants of the complexes $M_1 L_2$ and $M_2 L_1$ are made small enough compared with those of $M_1 L_1$ and $M_2 L_2$ so that there is no competition of M_2 for M_1 sites and vice-versa, unless there are excesses of the metals over their respective ligands.

The 'conditional' stability constants are related to the conventional stoichiometric stability constants by

$$\text{Log } K'_{M_1 L_1} = \text{Log } K_{M_1 L_1} - \text{Log } \alpha_{M_1} - \text{Log } \alpha_{L_1}$$

where

$$\alpha_{M_1} = 1 + \sum_{i=1}^{N} \beta_i [L_2]^i$$

$$\alpha_{L_1} = 1 + \sum_{i=1}^{P} \beta_i^H [H^+]^i + K_{M_2 L_1} [M_2]$$

The β_i^H, are the overall formation constants of the species LH_i, $K_{M_2 L_1}$ the stability constant of the complex formed between M_2 and L_1 and the β_i are the overall formation constants of any $M_1 L_2$ complexes formed.

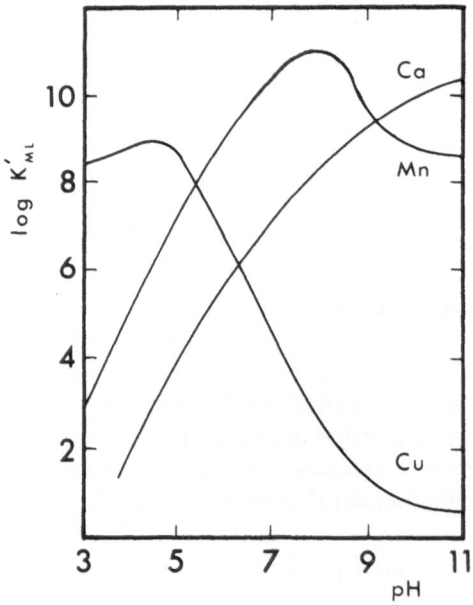

Fig. 5. Variation of the conditional stability constants of the Ca^{2+}, Mn^{2+} and Cu^{2+} complexes of EDTA with *pH* in the presence of 0.1 M triaminotriethylamine.

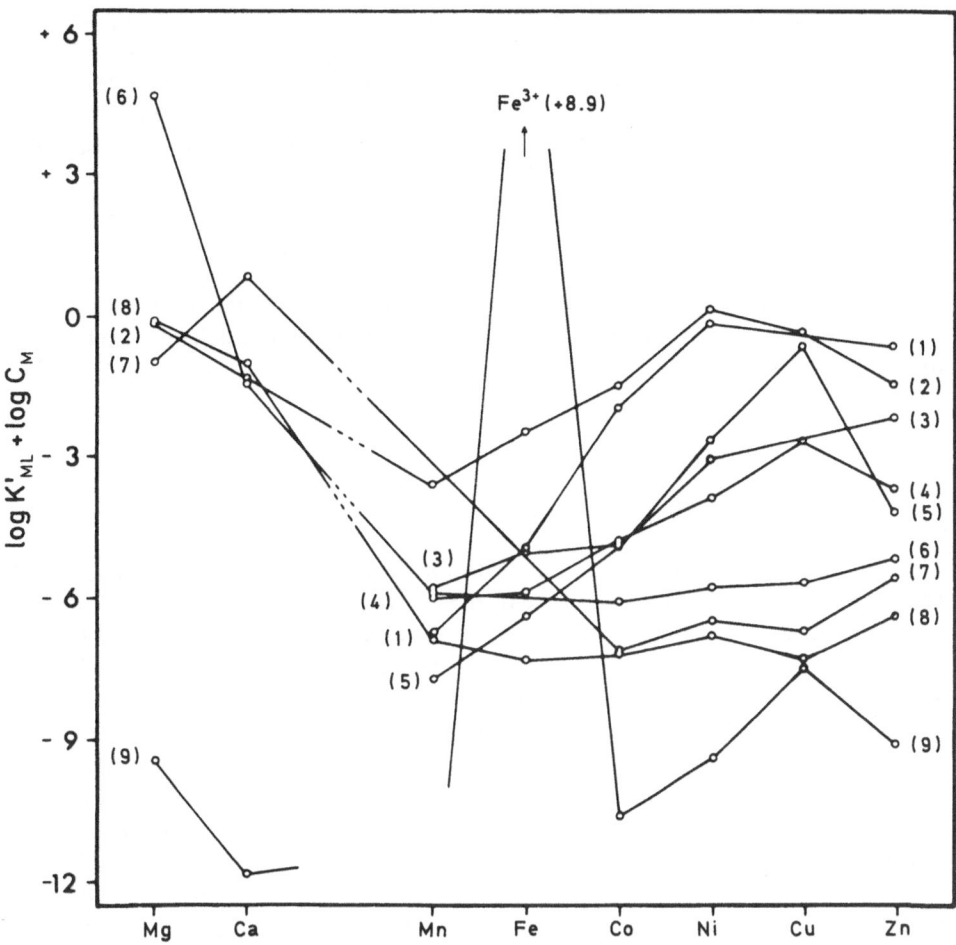

Fig. 6. Values of log $(K'_{ML} \cdot C_M)$ for a series of divalent metal ions and several types of ligands, assuming that these are present in a solution with the inorganic composition of sea water, at $pH = 8$ and $T = 25\,^\circ C$ (W. Stumm and P. A. Brauner in "Chemical Oceanography", ed. J.P. Riley and G.B. Kirrow, Academic Press, London 1975).

Ligands = (1) Cysteine; (2) O-Phenanthroline; (3) Thioglycollic acid; (4) Glycine; (5) Ethylenediammine; (6) Triphosphate; (7) Oxydiacetic acid; (8) Acetic acid; (9) Desferrioxamine B.

There are well known examples of this type of 'selection' in analytical chemistry; calcium and manganese, for instance, can be titrated by EDTA in the presence of copper if triaminotriethylamine ("tren") is added to the medium. The nett effect is a reduction of the conditional stability constant of the copper-EDTA complex by competition with "tren", as shown in Fig. 5. Once again, the product $K'_{ML} \cdot C_M$ (not the free concentration of the metal since this is taken into account in the value of K'_{ML}) gives a measure of the potential uptake capability of each metal ion by different types of ligands as shown in Fig. 6. It is clearly seen that Mg^{2+} is taken up preferentially by phosphate ligands, Ca^{2+} by polycarboxylate ligands, Fe as Fe^{3+} by polyphenolates or as Fe^{2+} by aromatic nitrogen ligands, Cu^{2+} by aminoacids or aliphatic nitrogen ligands and Zn^{2+} by sulphide ligands. Ni^{2+} is taken up by aromatic nitrogen and by sulphides, but the pattern relative to copper(II) is mainly the result of the higher free concentration of that metal in sea water.

Biology can, of course, make use of similar effects to select one metal cation from others competing for the same site.

Some other aspects are best dealt with discussing some concrete examples, to stress the variety of possibilities offered for the solution of the problems posed by uptake and selection.

4.1. Uptake of Magnesium

There is as yet no known biochemical device for the uptake of magnesium. Magnesium has presented special problems because of the analytical problems it poses, but there is no doubt however that such devices exist. The uptake of magnesium into *E. Coli* occurs from a concentration of 10^{-6} M providing an internal cell concentration of $\geqslant 10^{-3} M$. This energised uptake requires a binding ligand which gives a binding constant of $\log K_M \geqslant 6$. The system must not take calcium and must therefore have an effective binding constant for calcium $\log K_M \simeq 3$ and it must not take transition metal ions. There are two ways of developing a calcium free system: the first utilises an appropriate choice of liganding atoms; thus nitrogen, enolate or phenolate ligands bind magnesium more strongly than calcium; the second utilises the large size difference between the two metal ions. In developing such a suitable uptake site however great care must be taken to avoid binding of other metal ions, especially those from the transition series; for example, in chlorophyll synthesis in vitro it is easy to insert contaminating zinc into chlorin. Thus it would appear that only one (or two?) nitrogen centres but certainly not four could be used in a Mg^{2+} — scavenging ligand. If we combine both the possibilities discussed above, we can devise a suitable hole made from imidazole, carboxylate and possible phenolate which we can write as (N, O^-, O^-). If we take the product $K_M [M]$, Mg^{2+} is selected against Ca^{2+} by virtue of K_M but is selected against zinc by virtue of relative $[M]$ despite a rather adverse relative K_M.

We stress that as Mg^{2+} in cells and in body fluids is $10^{-3} M$, the best uptake sites for magnesium known in higher animal systems and many plants do not need to have binding constants much greater than $\log K_M = 4$. Undoubtedly chlorophyll binds more strongly, but the problem of the insertion of magnesium into chlorophyll is still a mystery; it is achieved in an enzymic process which must exclude other metal ions at a stage before the insertion step. Thus, there must be positive kinetic guidance (selectivity), after the initial uptake, see below.

4.2. Rejection of Calcium

Although there is little need to reject most metal ions, it is required to lower the concentration of both sodium and calcium in cells. Thus, internal sites in membranes must be available at which $\log K_{Ca} > \log K_{Mg}$ by a factor of 10^5 ($[Mg]_{in} = 10^{-3}M$; $[Ca]_{in} = 10^{-7}$) and $\log K_{Na} > \log K_K$ by a factor of 10^2 ($[Na]_{in} = 10^{-2}$; $[K]_{in} = 10^{-1}$). The latter condition can be satisfied by the choice of cyclic ligand while the former can be achieved by an appropriate choice of multi-carboxylate ligand, Table 10, of appropriate stereochemical restriction.

4.3. Uptake of Zinc

In the above sections we have shown that the product $\log K_M \cdot [M]$ for a given ligand and metal ion is a useful measure of the probability of uptake. Use of the product showed that certain metal ions can be selectively bound: calcium by four carboxylates in a protein; magnesium probably by a histidine and two carboxylates in proteins; ferric iron by enolate or phenolate oxygen ligands. The divalent transition metal ions bind too weakly to these ligands, for their product $\log K_M [M]$ is too low to give bind-

Table 10. Comparison of Ca^{2+} and Mg^{2+} binding constants to several ligands ($\log K_M$) T = 20 °C or 25 °C

Ligand	Binding constants		Ionic strength	$\Delta \log \dfrac{K_{Ca}}{K_{Mg}}$
	Mg^{2+}	Ca^{2+}		
Acetate	1.25	1.24	0 corr.	− 0.01
Malonate	1.95	1.85	0.1	− 0.10
Citrate	3.40	3.55	0.1	+ 0.15
Phosphate (HPO_4^{2-})	1.60	1.33	0.1	− 0.3
Pyrophosphate ($P_2O_7^{4-}$)	5.4	4.9	1	− 0.5
Triphosphate ($P_3O_{10}^{5-}$)	7.1	6.3	0.1	− 0.8
Glycinate	3.4	1.4	0	− 2.0
Imidodiacetate	2.94	2.59	0.1	− 0.35
Nitrilotriacetate	5.36	6.57	0.1	+ 1.2
Uramildiacetate	8.2	8.3	0.1	+ 0.1
EDTA	8.6	10.5	0.1	+ 1.9
EGTA	5.4	10.7	0.1	+ 5.3
Picolinate	2.23	1.8	0.1	− 0.4
Dipicolinate	2.30	4.40	0.1	+ 2.1

Source: Stability constants. The Chemical Society Sp. Publ. No. 17 and 25 London, 1964 and 1971.

ing. Binding of the same metal ions to a group of two, three or four *nitrogen* or *sulphur* donors is much stronger and it is observed that a metal such as zinc is taken up by (RS^-, RS^-, RS^-, RS^-), (N, N, O^-) or (N, N, RS^-) in different proteins. The question now arises as to how zinc can be bound in such sites to the exclusion of other transition metal ions such as copper(II) and iron(III). Below, we shall show that there is a preferred site for Cu(II) which may so lower the free [Cu(II)] that $\log K_M [M]$ for copper is small in comparison with $\log K_M [M]$ for zinc for the sites just listed. As Zn(II) is the best available electron acceptor after copper(II) as judged by the product $\log K_M [M]$, it then follows that Zn(II) will occupy these sites. However, it is found that in some proteins, e.g. rubredoxin, Fe(III) binds in a site of precisely the same four donors (RS^-, RS^-, RS^-, RS^-) as that found for Zn(II) in its specific proteins, e.g. alcohol dehydrogenase. Moreover crystallographic studies demonstrate that the two sites are of the same approximately tetrahedral symmetry. If we assume that the uptake of the two metal ions into the two different sites is due to thermodynamic factors then we must presume that either there are structural features in the second coordination sphere which establishes the different selective reaction or that the cavity made by four RS^- groups can not be of the same dimensions in the two proteins. In the latter case, the selection between Fe(III) and Zn(II) can be compared with that between Mg(II) and

Table 11. Stability Constants ($\log K_M$) of complexes of VO^{2+} with Several Types of Ligands

Ligand	Log K_M	Temperature °C	Ionic Strength
Acetylacetone	8.68	25	–
Catechol	15.20	30	0.1
Citric Acid	8.83	20	1
DCTA	19.40	20	0.1
EDTA	18.77	20	0.1
8-Hydroxyquinoline	10.79	25	0.1
8-Hydroquinoline-5-Sulphonic Acid	11.79	25	0.1
Oxalic Acid	9.75	–	0.05
1.10-Phenanthroline	5.47	25	0.1
Salicyllic Acid	13.38	25	0.1
Sulphosalicyllic Acid	11.71	25	0.1
Thiazolylazoresorcinol	11.2	25	0.1
Tiron (4.5-Dihydroxybenzene – 1.3-Disulphonic Acid)	16.74	25	0.1

Source: Stability constants. The Chemical Society Special Publ. No. 17 and No. 25 (Supplement), London 1964 and 1971.

Na(I) in the ring chelates (see section III.3.2.) for preference would depend upon a precise matching of cavity size to metal ion size. It can be seen that through the operation of several factors — the concentration of free ions, ligand electronegativity, cation and ligand charge and of cavity size — the uptake of metals is potentially quite selective. However, we shall see that further factors can be introduced based on stereochemistry which increase the selectivity still further.

4.4. Uptake of Vanadium

The uptake of vanadium is a very special case known in detail for the sea cucumber only. Here vanadium is found as VO^{2+} in association with a special complexing ligand and in a medium of about $2\,M\,H_2SO_4$. The choice of the acid medium could be due to the requirement to control the redox properties of vanadium or to provide a selective medium for chelation. VO^{2+} has a high affinity for di-phenolates or for ligands of the type 8-hydroxyquinoline, Table 11. Here we see how the design of biology descends even to the control of the medium, so that selective accummulation can be attained, see Fig. 3.

5. Effect of Preferential Coordination Geometry

5.1. Zinc and Copper

Zinc and copper uptake illustrate other features of selectivity. Both cations are only available in small amounts, though zinc may be 100 times more readily available in soils and water than is copper. To counterbalance this, copper binds to most ligand centres at least 100 times more strongly than does zinc. Thus $\log K_M [M]$ is similar for both metals with many ligands. How can selective uptake be achieved with 99% efficiency as is required in *both* copper and zinc enzymes? We believe that here the selectivity of biology is based upon their different polarising power and the different stereochemical demands of the cations. In the first place copper(II) shows a remarkable ability not shared with zinc to form a bond·at $pH = 7$ to peptide nitrogen in its ionised state; in this case the selectivity factor copper/zinc is perhaps as great as 10^4. Positive selection for copper is further enhanced by utilising the preference of copper for a tetragonal hole which arises from the Jahn-Teller effect. The copper carrier protein of blood utilises these factors for one or two of the metal-nitrogen bonds are ionised amides while the geometry must be a roughly square co-planar four nitrogen ligand chelate. Sakar and his colleagues (12) have designed ligands (peptides) based on the terminal sequence of this carrier protein and have shown that such peptides may well be valuable drugs assisting redistribution of copper in patients defect of copper metabolism. Provided all copper is removed by such ligands, which must be present in excess over copper, there is little competition for zinc.

It was pointed out long ago, however, that positive selectivity for zinc could be based on its different demand, now for tetrahedral geometry, with nitrogen or sulphur donor ligands (13). The ideal zinc site will then be any combination of four ligands formed from nitrogen or sulphur donors in tetrahedral array (see Table 12). Many instances of this or a close parallel to it, are now known in biology amongst structurally important zinc sites, while enzymic sites where zinc is "entatic" are somewhat distorted from this geometry.

A fine example of zinc and copper selectivity appears in superoxide dismutase (14). The zinc site (not active) is close to tetrahedral and the copper site (active) is a slightly distorted tetragonal hole. Both holes are made from nitrogen bases one of which is an ionised anionic nitrogen of imidazole. A parallel finding is that the site for zinc in myoglobin differs from the site for copper (15). The difference between the sites is related to the above difference in superoxide dismutase. We note that the requirement for selectivity of uptake works contrary to the demand to activate the metal ion site (entatic state) and undoubtedly biological binding sites have evolved to meet both requirements.

Table 12. Some Zn^{2+} enzymes and proteins

Zinc Protein	Source	Ligands	Geometry of site
Carboxypeptidase	bovine pancreas	2 his; 1 glu; 1 w	dist. tetrahedral
Thermolysin	B. thermolyticum	2 his; 1 glu; 1 w	dist. tetrahedral
Alcohol dehydrogenase	Horse liver (2 sites)	4 cys: 2 cys; 1 his; 1w	tetrahedral dist. tetrahedral
Carbonic anhydrase	Human	3 his; 1 w	dist. tetrahedral
Insulin	porcine pancreas	3 his; 3 w	trigonal field
Cu^{2+}/Zn^{2+} Superoxide dismutase	see Table 13	3 his; 1 asp	tetrahedral

5.2. Manganese (III) and Iron (III)

There is a second case in the first row transition metals where strong advantage can be taken of the contribution to stability of the Jahn-Teller distortions associated with the $3d$ subshell, that of Mn^{3+}. Thus, we predict that the uptake of Mn^{3+} rather than Fe^{3+} into proteins will be based on the distinction between tetragonal (Mn) and tetrahedral, or octahedral (Fe) geometries. Of course, tyrosines must be favoured ligands in these cases of highly charged cations; one example is that of the different superoxide dismutases (see Table 13), and another could be that of the O_2 releasing protein of photosynthesis (16). Thus we can make a Table of selectivity (Table 14) in which we list the preference of transition metal ions for different sites. The best nitrogen donor at $pH = 7$ would be histidine, but it is found that this is a rather non-discriminating

Table 13. Superoxide Dismutases

Type	Source	Approx. Molecular weight	Number of Sub-units	g. ion of metals per sub-unit
I Blue-green Cu^{2+}/Zn^{2+}	Bovine heart Bovine erythrocytes Green Pea Neurospora Crassa Saccharamices Cerevisiae Photobacterium Leiognathi	~ 32,000	2	$1\ Cu^{2+}$ $1\ Zn^{2+}$
II Wine-red Mn^{3+}	Escherichia Coli B Streptococus Mutans	~ 40,000	2 2	$1\ Mn^{3+}$

	Pleurotus oliarius
	  (mitochondria?)
	Bovine liver mitochondria
	Chicken liver mitochondria
	Human liver mitochondria
	Isoenzyme B from human
	  tissues
		~ 80,000	
			
			4
			4
			4	

III Yellow Fe^{3+}	Escherichia Coli B (periplasmic space) Plectonema Boryanum Spirulina Platensis Photobacterium sepia Photobacterium leiognathi	~ 38,000	2 2	 $1\ Fe^{3+}$

Source: *Fridovich, I;* Ann. Rev. Biochemistry, *44*, 147, (1975)

general ligand which is used in binding Mg(II), Mn(II), Cu(II), Zn(II), Fe(II) and even possibly Co(III) in B_{12} carrier proteins. At $pH = 7$ we can design ligands selective for all four metals using RS^-/histidine combinations. Finally, we note that given all these factors it is possible to conceive of circumstances in which $\log K_M [M]$ would result in the uptake of Ni(II) or Cd(II) despite their low availabilities.

Table 14. Table of selectivity

	Geometry of site		Ligand atoms		
	Tetragonal	Tetrahedral	Phenolate (O^-)	Ionised peptide (N^-)	Thiolate (RS^-)
Cu(II)	+			+	?
Fe(III)		+	+	?	+
Mn(III)	+		+	?	?
Zn(II)		+			+

6. Uptake of Transition Metals in Preformed Rings

The selective scavenging we have discussed above can have its selectivity introduced as the protein folds about the metal ion, for the different stereochemistries and sizes of M_1 and M_2 can produce strain energy which rejects M_2 while M_1 is accepted.

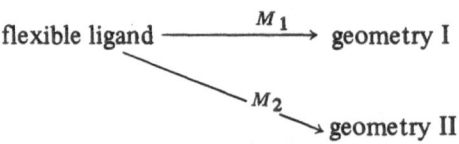

Another similar device in selection may depend upon a ligand made to fit exactly to a given ion size, as was described for Group II A metal ions, but with transition metals the effect can be more subtle. Imagine a high-spin cation approaching a rigid hole generated by a group of ligands; if at close approach the ion is able to change spin-state with this group of ligands and can, on so doing, exactly fit the hole in the ligand, it gains a specific energy increment. Thus advantage accrues from a combination of a ligand which gives a correct geometry and electronegativity to assist a spin-state change of a particular cation utilising a hole of a precise size. Consider four cations, Fe^{2+}, Co^{2+}, Cu^{2+}, Zn^{2+}, approaching a tetragonal hole. At long distances size differences are small and possibly bonding is favoured in the order of their electron affinity: $Cu > Zn > Co > Fe$. Now, as the metals approach, spin-state changes are favoured following ease of spin-pairing which are $Co > Fe$, and $Cu, Zn = 0$; thus this ligand could allow low-spin cobalt to become preferred. Many features of this reaction are parallel to those found in the selective analytical reagents for iron (II) and copper (I). The ligand of choice for iron (II) is orthophenanthroline which permits a spin-state change (see Fig. 7) while the ligand of choice for copper (I) is a sterically hindered 2-substituted orthophenanthroline which does not give a spin-state change with iron (II). We believe this is the principle behind vitamin B_{12} selection for cobalt (II) and further stability results as in the site the Co (II) becomes Co (III), thus achieving great kinetic stability. We return to the cases of the uptake of iron, copper and magnesium in a later section.

7. Cooperativity at Multi-Metal Ion Centres

Binding of metal ions in pairs or even four at a time is well recognised in chemistry and now in proteins. While configurational entropic considerations oppose such polymerisations, the cooperativity of cation/anion systems which is seen most markedly in ionic salt lattices such as NaCl and ZnS operates strongly in support of polymer formation.

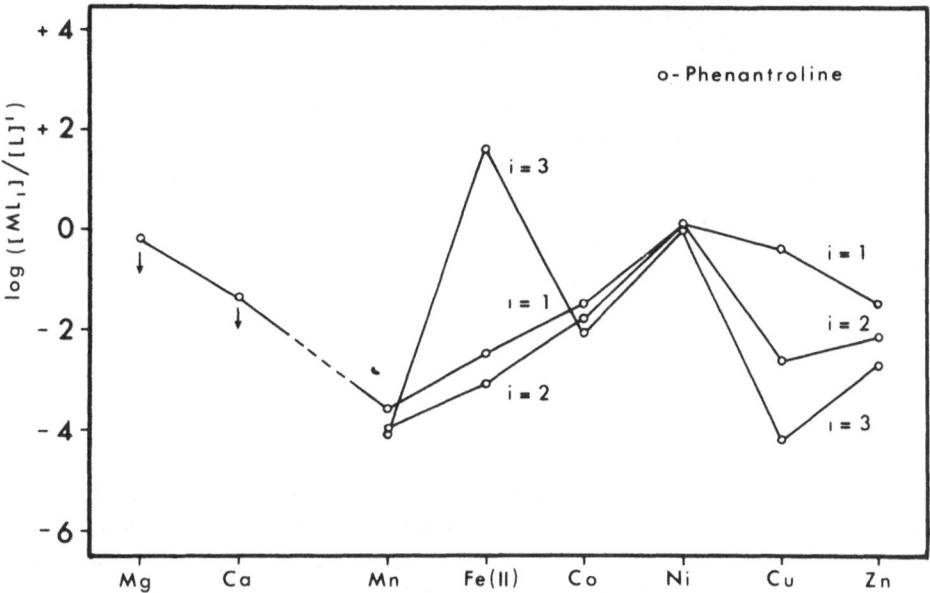

Fig. 7. Effect of spin-pairing in the Fe (II)-trisphenanthroline complex forming in sea-water, giving a more favourable

$$[ML_i]/[L]^i = K'_{ML_3} \cdot C_M$$

ratio for the 3 : 1 than for 1 : 1 and 2 : 1 complexes. The figure also illustrates the effect of steric hindrance in the Cu (II) 2 : 1 and 3 : 1 complexes. Note that the relative positions of the values for the different ions is affected by the concentration of the metal ions in the medium.

In fact such polymerisation is found to occur especially with the smaller highly charged ligands e.g. O^{2-}, OH^-, S^{2-}, and RCO_2^-. The opposed entropy term is greatly reduced if the anionic centres are already built into a preformer such as a protein and is still further reduced if the protein is largely folded into the correct conformation for metal ion uptake. A simple example is the uptake of Fe(II) into hemerythrin where there are two metal sites close together but, as shown by many physical properties, not into a true polymer. Oxidation to Fe(III) produces a true polymerisation in which either an O^{2-} or $-CO_2^-$ is shared as a bridging group. It could be that this binding site is based on a similarity with the size of $[Fe_2(OH)_2]^{4+}$ which is well-known in aqueous media. In copper/zinc superoxide dismutase the shared protein ligand is (imidazole)$^-$, an anion which could not exist but for the fact that it is sandwiched between two metals, Zn and Cu, and is therefore held cooperatively (17). Further examples are the Fe_2 and Fe_4 units of ferredoxins, where S^{2-} acts as a bridge and cysteinate as a protein ligand (18) and the Cu_2 dimers in many proteins where the bridge is as yet unknown but could be disulphide (*19*). The bridging anion can also be O_2^{2-} when either a $(Cu^I)_2$ or a $(Fe^{II})_2$ dimer picks up oxygen to give $(Cu^{II})_2$ and $(Fe^{(III)})_2$ dimers respectively (*20*).

We may now ask about selectivity in such processes. Obviously polymerisation (formation of crystals) of inorganic compounds is highly selective and "foreign" atoms are strongly excluded from the lattice as they cause unfavourable distortions. In general there is a preference in polymerisations for ions of precisely the same size — i.e. homo-polymers are favoured. It could be that this selectivity is used by biology taking up the Fe_4S_4 clusters, and not one iron atom at a time.

8. *Kinetic Control of Metal Chelate Formation*

In the above we have explicitly or implicitly assumed that the incorporation of metal ions into specific sites was governed by thermodynamic factors. As stated in the introduction this is a feature of the initial uptake step, but a further step can occur and it may virtually be an irreversible one for a particular metal:

$$M + L_1 \underset{\text{initial} \atop \text{uptake}}{\xrightarrow{K_{ML_1}}} ML_1 \underset{+L_2}{\xrightarrow{k}} \underset{\text{final} \atop \text{state}}{ML_2} + L_1$$

For example, there is good reason to believe that the magnesium of chlorophyll is inserted into the chlorin ring and it could be that iron and cobalt are inserted into porphyrin and corrin respectively. Such reactions can be of much greater selectivity that the initial uptake at equilibrium, for control can rest largely in the rate determining step. For the initial uptake step as written above we are interested in a ratio of rate constants for the various metal ions which may not vary greatly from one metal ion to another or might be unfavourable to a metal M_1 as opposed to another M_2. It could well be that, for certain purposes, selectivity in this step was deliberately avoided (see below). Let us now suppose that both metals are in fast exchange between states as free ions and states as ML_1; this state is then determined by the different values of $K_{ML_1} \cdot [M]$. The order of introduction of the metals into L_2 now becomes dependent on $k \cdot K_{ML_1}$, but $[M]$ and the rates of transfer k, from ML_1 to ML_2, could be very different from the stability orders K_{ML_1} or from stability orders K_{ML_2}. In fact L_1 could merely act to control this rate.

An example makes clear the very high selectivity which can thus be introduced. The chelation of magnesium by the protein which inserts Mg^{2+} into chlorophyll can not have a binding constant for Mg^{2+} which is higher that that for Zn^{2+} for we have never made such a ligand; on the contrary, it is likely that $\log K_{ZnL_1}$ will be orders of magnitude higher than $\log K_{MgL_1}$. Nevertheless, since the concentrations $[Mg^{2+}]$ and $[Zn^{2+}]$, are vastly different in biological systems, the protein may be bound, say some

1%, by Zn^{2+} and 99% by Mg^{2+}, although this will only happen with a relatively poor ligand e.g. an (N, O^-, O^-) centre, see earlier. Now, the very nature of chlorophyll, a porphyrin-like ligand with four nitrogen centres, would lead us to suppose that the $\log K_{porphyrin.Zn}$ would exceed $\log K_{porphyrin.Mg}$ by at least ten units. Thus, at equilibrium, it is inevitable that zinc chlorophyll would be formed after transfer from ML_1 and not magnesium chlorophyll. As chlorophyll must be such a good ligand for zinc it is hard to see how any reasonably devised step could prevent zinc rather than magnesium from entering chlorophyll at *equilibrium* because of the huge bias in the chlorophyll binding constants. Now let us suppose that the step k_{ML_2} is irreversible. It is clear that ML_1 can be such that the rate for magnesium $ML_1 \to ML_2$ could be many orders of magnitude faster than for Zn^{2+} *(21)*. This could arise in part as ML_1 contains 99% Mg^{2+} and only 1% Zn^{2+} but also because MgL_1 (not ZnL_1) was recognised by L_2; as a consequence magnesium could form ML_2 *and*, as the step is irreversible, this fast reaction would *block* the entrance of zinc. Thus, those weak-binding metal ions which are *in excess* over ligands but react rapidly can block the entry of ions which bind better so long as the kinetics are suitable arranged. It is a striking observation that while *in vivo* the insertion is performed correctly giving magnesium chlorophyll, *in vitro* it is very difficult to avoid the production of zinc chlorin when using disrupted enzyme systems in an effort to make magnesium chlorin.

In fact, it seems likely that the correct insertion of each metal ion with less than 0.1% error is achieved for magnesium in chlorin, iron in protoporphyrin, cobalt in corrin and copper in uroporphyrin; probably this result could only be brought about by kinetic controls. Such purity of product is of course essential; one iron chlorin in four hundred chlorophyll molecules could not be permitted, for this would introduce an electron trap into a required photo-process insulator. As far as the authors are aware this error has not been observed, but it could be a limitation on the number of chlorophyll units in the light-harvesting device. Be this as it may, it appears that only one chelating agent is used for one selected metal ion to within the limits of analytical detection. A minor exception is that of the appearance of some zinc porphyrin in certain blood diseases.

Using several kinetic devices biological systems have managed to match ligands with metal ions as follows:

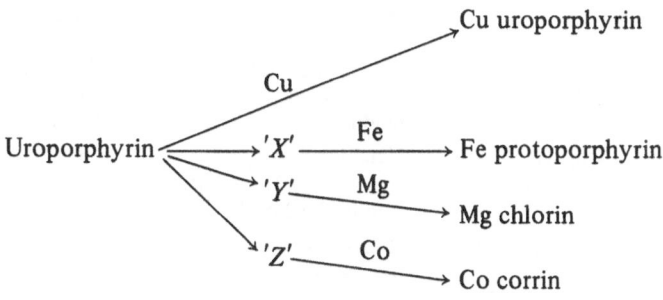

where $'X'$ $'Y'$ $'Z'$ are intermediates in synthesis, the ML_1 kinetic devices.

9. Summary: Uptake and Retention of Metal Ions

While it is known that certain enzymes (chelatases) are available for the insertion under kinetic control of different metal ions into the above complicated and special chelating agents, this cannot be the general method of synthetic control. Rather it seems highly likely that the common situation in metallo-proteins is that the uptake of the metal is under thermodynamic control. Here two cases must be distinguished, see above: (1) Weak binding by the protein for all metal ions where the ligand is present in smaller quantities than the total number of metal ions which can bind; these are the centres for metal ions such as K^+, Na^+, Mg^{2+} and Ca^{2+}. (2) Sites which cannot bind Na^+, K^+, Mg^{2+} and Ca^{2+} and in which there are binding sites of nitrogen and sulphur ligands of chosen geometry; these are the sites for iron, cobalt, nickel, copper and zinc. These proteins are produced in smaller amounts.

In the latter case the question arises as to how all the sites for divalent cations are not filled by copper (II) ions as the stability series for nitrogen and sulphur ligands is

$$Cu^{2+} \gg Ni^{2+}, Zn^{2+} > Co^{2+} > Fe^{2+} \ (> Mn^{2+})$$

In the absence of spin-state and valence state changes, the difference across the series for a four nitrogen centre donor protein can be as high as 10^{10}, which would exclude Fe^{2+} binding. This factor is reduced somewhat by considerations of availability which may roughly counterbalance the difference between Cu^{2+} and Zn^{2+}, so that there could well be both Cu^{2+} and Zn^{2+} selective proteins, but, on the basis of model studies, Ni^{2+} and especially Fe^{2+} cannot have a product K_{ML} [M] higher than that for Cu^{2+} for any such nitrogen/sulphur centre. The only way in which selectivity can then operate is by removing Cu^{2+} first and highly selectively from the system into special complexes, so that $[Cu^{2+}]$ is made very small indeed. Hence for copper and perhaps zinc there must be a slight excess of dominant ligands. Selectivity can then be based on the following table, (Table 15).

Table 15. Classification of Metal Ions

Very good acceptors		Moderate acceptors		Poor acceptors	
Cu^{2+}	Zn^{2+}	Fe^{2+}	Mn^{2+}	Mg^{2+}	Ca^{2+}
Specifically bound at very strong sites		Intermediate binding		General weak binding centre	
Ligand in (slight) excess		Metal ion and ligand roughly equal; possible feed back control		Metal ions in excess	
No free ions in solution		Modest concentration of free cations		High concentrations of free ions	

10. The Blood Stream as an Example

The blood stream must distribute the metal ions and Table 16 shows the known carriers it contains. The system operates as described above: copper is removed by a very powerful chelating agent using an ionised peptide, which is a very special Cu(II) ligand; iron is oxidised to Fe(III) and removed to storage as Fe.O.OH in ferritin, for the solubility product of Fe(III) hydroxide is 10^{-36}. From the store, the blood-protein transferrin picks up the Fe(III) using phenolate binding groups (note the use of OH^-, O^{2-}, and phenolate$^-$ (small anions) to control Fe(III)). Zinc is also removed by a zinc albumin, but this is a little known protein; we suspect that it has three or four N/S donors and gives a tetrahedral geometry.

At the other end of the scale Na^+, K^+, Mg^{2+}, and, to some degree, Ca^{2+}, are free in the blood solution. When extra calcium is required it cannot travel as the free ion, as this increase would exceed the solubility product of calcium phosphate or carbonate; there is then a special calcium carrier protein which is generated under hormone control and which has phosphorylated side chains. The removal of all the transition metal ions such as Fe(III) and Cu(II) by the strong binding centres described above prevents competition by them for these Ca(II) sites. There are other carrier proteins for vitamin B_{12} (Co), Mn, etc.

The transport of the different ions by different proteins allows specific recognition of the metal ion *through* the protein at given cell surfaces

$$M + L \rightleftharpoons ML_1 \xrightarrow{L_2} ML_2 + L_1$$

and thus the ions, especially iron to erythrocytes and calcium to bone cells, can be delivered to appropriate sites under kinetic control.

Table 16. Carriers in Blood Stream

Metal Ion	Free [M]	Carrier	Basis of Selectivity
Iron (and Manganese?)	10^{-10}	Conalbumin Transferrin	Phenolates bind M^{3+}
Copper	10^{-10}	An albumin	Ionised peptide N^- binds copper
Calcium	10^{-3}	Phospho-proteins	Weak binding for most metals
Zinc	10^{-8}	An albumin	Not known
Mg^{2+}, Na^+, K^+, (Ca^{2+})	10^{-3} to 10^{-1}	None	Freely circulating as ions
Co^{3+} (vitamin B_{12})	10^{-10}	Carrier protein	The B_{12} side-chains are recognised.

IV. Uptake of Neutral Molecules

A few elements are taken up as neutral molecules i.e. not as anions or cations. The most notable are elements which can be taken from the air, O_2, N_2 and CO_2. It appears that CO_2 is converted rapidly to bicarbonate which is absorbed at an anion binding centre, carbonic anhydrase, or that it reacts directly with an activated organic molecule so that the CO_2 incorporation is energy driven. In the last case CO_2 is caught immediately in a kinetic trap. Carbon dioxide has a large quadrupole and is very different from O_2 and N_2.

Oxygen, O_2, and nitrogen, N_2, are both absorbed at metal centres. For O_2 several different types of centre are known in hemoglobin (Fe(II)), hemerythrin (Fe(II) ... Fe(II)) and haemocyanin (Cu(I) ... Cu(I)). In the last two cases the oxygen is held in the form O_2^{2-}, but the electronic state of O_2 in hemoglobin is still controversial (20). Final capture of the element oxygen involves the attack of O_2 activated at a metallo-enzyme centre to give organic oxygen containing molecules.

The uptake of N_2 is known to be by a special kinetic trap, the enzyme nitrogenase, which carries out the reaction

$$6e + N_2 + 6\,H_2O \rightarrow 2\,NH_3 + 6\,OH^-$$

The resulting ammonia goes on to become incorporated in organic molecules by condensation reactions. The nitrogenase reaction requires ATP (energy) and is thought to be *irreversible*; thus the reaction is very like the uptake of CO_2. It is immediately clear that O_2 and N_2 are handled in very different ways and by very different uptake centres. It is not quite certain which atomic grouping picks up N_2 but a reasonable conjecture is that the group is of two molybdenum atoms (20). How is it that this group does not pick up O_2? The reverse question of why N_2 is not picked up by O_2-absorbing sites is readily answered, for N_2 has but little affinity for most metal ion binding sites; N_2 is an inert molecule compared with O_2, for O_2 has a greater electron affinity and is a diradical.

Turning to inorganic chemistry there is only one certain case where N_2 is absorbed in preference to O_2, the reaction (22):

$$2\,[Ru(NH_3)_5\,H_2O]^{2+} + N_2 \rightleftarrows [Ru(NH_3)_5 \cdot N_2 \cdot Ru(NH_3)_5]^{4+} + 2\,H_2O$$

To the best of the authors' knowledge all other N_2-uptake reactions must be carried out in the absence of O_2. Of course it is possible that biology can prevent access of O_2 to the N_2 reaction site, but this is unlikely as O_2 can diffuse rapidly *through* proteins and membranes. It is just possible that N_2 is picked up preferentially by a good σ-acceptor centre.

V. The Uptake of Non-Metals

1. General Considerations

Before turning to the processes of non-metal uptake some simple chemical points are worth note. Non-metal elements do not form simple *atomic* anions of any stability except in Group VII: F^-, Cl^-, Br^-, I^-. There is therefore little possibility of direct competition between non-metals on the basis of similar atomic size and charge for even these anions have very different radii, see Table 17. Only the anions, F^- and OH^-, and Br^- and SH^-, could show such competition, but OH^- and SH^- are of low availability at $pH = 7$. The usual forms in which the non-metals of Group IV, V and VI occur are as oxo-anions e.g. CO_3^{2-} (HCO_3^-), SiO_4^{4-}, NO_3^-, PO_4^{3-} ($PO_4H_2^-$, PO_4H^{2-}), AsO_4^{3-}, SO_4^{2-}, SeO_4^{2-}, or polymerised forms of these anions. More reduced states of these elements are not thermodynamically favoured at $pH = 7$ in water and in air. The protonation of the anions can change their charge without changing appreciably their size or shape; the maximum change is shown by carbonate on going to bicarbonate when the radius of the CO_3^{2-} disc decreases from 1.85 A° to at most 1.63 A°. The sizes of more complex anions treated as discs or spheres (thermochemical radii) are given

Table 17. Pauling's crystal radii of common anions

Ion	Radius (A)	Ion	Radius (A)	Ion	Radius (A)
F^-	1.36	O^{2-}	1.40	OH^-	1.40
Cl^-	1.81	S^{2-}	1.84	SH^-	1.95
Br^-	1.95	Se^{2-}	1.98		
I^-	2.16	Te^{2-}	2.21		

Pauling, L.: The nature of the Chemical Bond, 3rd. ed. Cornell Univ. Press, Ithaca, N.Y., 1960.

Table 18. Thermochemical radii of common anions

Ion	Radius (A)	Ion	Radius (A)	Ion	Radius (A)	Ion	Radius (A)
CO_3^{2-}	1.85	NO_3^-	1.89	OH^-	1.40		
HCO_3^-	1.63						
SiO_4^{2-}	2.40	PO_4^{3-}	2.38	SO_4^{2-}	2.30	ClO_4^-	2.36
		AsO_4^{3-}	2.48	SeO_4^{2-}	2.43		
BO_3^{3-}	1.91						
MoO_4^{2-}	2.54						
		SbO_4^{3-}	2.60	TeO_4^{2-}	2.54	IO_4^-	2.49

Waddington, T.C.: Adv. Inorg. Chem. Radiochem. Vol. 1, p. 180. London: Academic Press 1959.

in Table 18. In these cases there could well be competition for uptake between some of the oxyanions.

We have seen that capture of a metal ion is largely dependent on appropriate complex formation; now, the binding in these complexes has a heavy ionic component and as a consequence this type of complex usually has kinetic lability. Thus, more or less complicated thermodynamic traps had to be designed for separation and storage of the cations largely without change of oxidation state as illustrated above. By contrast, changing the oxidation state of the non-metal anions or oxyanions makes them become capable of direct covalent attachment to the carbon skeletons of organic molecules, and thus they can be captured in kinetic traps as organic compounds.

Examples are

$I^- \rightarrow I^0 \rightarrow$ iodo compounds   (oxidation)

$NO_3^- \rightarrow NH_3 \rightarrow$ amide   (reduction)

$SO_4^= \rightarrow H_2S \rightarrow$ thiols   (reduction)

The only non-metals which can not easily be handled so as to make such kinetically stable covalently-bonded (to carbon) traps are F, (Cl), P, Si and B. Of these F and Cl are usually handled by biology in the same way as simple cations are handled i.e. uptake and binding as atomic ions, F^- and Cl^-. The other three are stable as high oxidation state oxo-anions such as HPO_4^{2-}, $H_2SiO_4^{2-}$, $H_2BO_3^-$, but, as these anions readily form condensed structures both with themselves and with organic hydroxy-groups, they can be trapped in semi-labile organic ester centres. (The semi-lability is due to the fairly ready back hydrolysis of their $X-O-X$ bonds.) Thus, virtually all non-metals can be caught and handled in *kinetic* traps. In fact the situation is very similar to that discussed in paragraph III.8 where we referred to the formation of such chelates as chlorophyll in a two-step process

$$M + L_1 \rightleftharpoons ML_1 \xrightarrow[L_2]{k} ML_2 + L_1 .$$

Here, the second step is the kinetic trap.

We see immediately that the problem of retaining the non-metal elements in biology is much easier than that of retaining the metal ions, but the initial problem of uptake is very similar. Hence, anions must be absorbed first on the basis of (1) size and charge, (2) bonding affinity for different types of cationic centres, e.g. H^+, NH_3^+, M^{n+} (compare ROH, RNH_2, RS^- for cations). The way in which they are absorbed, i.e. on the surface of cells, by scavenger chemicals etc., will not be listed here and it may be presumed that the modes of absorption are much the same as those already described for the cations. We pass directly then to the actual chemistry of the reversible binding of anions to different chemical groups.

2. Influence of Anion Size and Charge

The various oxyanions and anions, e.g. those of Group VII, show considerable variation in size, so F^-, Cl^-, Br^-, I^-, for example, should be readily separable by the correct design of absorbing holes and there is good reason to suppose that the associated thermodynamic selectivity can outweight the vast differences in abundance (availabilities) of these elements. However, the nature of the holes must be quite different from those which were found to be useful in separating cations as (1) anions are *negatively-charged*, (2) anions have a relatively lower charge density as they are relatively very large ($F^- \cong K^+$). The first point means that the orientation of binding dipoles of the agents for complexing to them must be in the opposite sense to that used in binding cations; hence, whereas a hole for a cation must be lined with R_2O, ROH, R_2S, RCO_2H or R_3N groups etc., that for an anion must be lined with a positive or neutral surface atom. Now, a curiosity of chemistry is that there are very few ways of getting the positive ends of dipoles exposed in polymeric substances of biological origin, for the positive ends, usually carbon, are frequently the backbone atoms. The exceptional case is $-H$ in $-OH$, $-NH$, $-SH$, and this type of positive dipolar centre often binds anions through H-bonds, though, commonly, at least one positively charged group must be present too. Another type of lining which is important in anion binding sites is provided by hydrophobic groups such as $-OCH_3$, $-SCH_3$, phenyl, $-CH_2-$, etc., despite the fact that they are of low polarity, but these hydrophobic centres will only take up anions if additionally they contain a *charged centre*. The simplest case is that in which the fold of a protein generates a positive charge which lies below the surface of the protein and is surrounded by non-polar side-chains. This type of "buried" charge is of greater importance in the uptake of anions than of cations just because the charge density of anions is so much lower; it is known that anions partition into organic solvents more easily than cations.

A deeper understanding of these observations depends upon a treatment of hydration of the anions, for the hydrated state is in equilibrium with the bound site, whether it is a hydrogen-bonding site or a hydrophobic site;

$$X^-(H_2O)_n + site^\oplus \rightleftarrows site^\ominus, \ X^- + nH_2O$$

This aspect is of considerable importance and it will be dealt with in a separate section.

3. The Hydration of Anions and Binding of Anions to Proteins

Values for the absolute ΔG_{hyd}, ΔH_{hyd}, and ΔS_{hyd} of hydration of the anions are given in Table 19. Relatively speaking, the entropy of hydration of large, lowly charged, anions is slightly unfavourable, whereas that of small, more highly charged, anions is very unfavourable to solution; the greater free energy of hydration ΔG_{hyd} of the small anions results from the much larger favourable heat of hydration ΔH_{hyd}. The simplest

Table 19. Selected values of absolute energies, enthalpies and entropies of hydration of anions at 25 °C[a]) and lyotropic numbers[b])

Anion	$-\Delta G^0_h$ Kcal mole^{-1}	$-\Delta H^0_h$ Kcal mole^{-1}	$-\Delta S^0_h$ e.u.	Lyotropic number
F^-	102	113	36	4.8
Cl^-	73	80	22	10
Br^-	66	72	19	11.3
I^-	59	63	13	12.5
OH^-	104	117; 117[d])	43	(4.2)[e])
ClO_3^-	67	75[c])	25	10.65
ClO_4^-	62	68[c])	19	11.8
BrO_3^-	75	83[c])	28	9.55
IO_3^-	93	103[c])	(36)[e])	6.25
NO_2^-	71	79[c]); 74[d])	27	10.2
NO_3^-	63	70; 68[d])	22	11.6
CN^-	70	78; 78[d])	28	(10.3)[e])
CNO^-	82	89	(23)[e])	(8.5)[e])
SCN^-	52	58[e])	(20)[e])	13.25
N_3^-	67	73	(22)[e])	(11.1)[e])
$HCOO^-$	86	95	(31)[e])	(2.6)[e])
CH_3COO^-	87	96	(31)[e])	(7.5)[e])
S^{2-}	295	309	47	—
CO_3^{2-}	298	321	76	—
SO_4^{2-}	235	254	63	—
SeO_4^{2-}	220	239	66	—
H^+	260.5	269.8	31.3	—

References:
[a]) *Vasil'ev et al.*: Russ. J. Phys. Chem. *34*, 840 (1960), from lattice enthalpies and entropies of solution, based on $\Delta H_f(Cs^+, g) = \Delta H_f(I^-, g)$, recalculated for $\Delta S^0_f(H^+, aq) = -5.3$ e.u. and $H_f(H, aq) = 95.8$ Kcal mole^{-1} (standard state H^+ (g, hyp. 1 atm) → H^+ (aq, hyp. 1 M))
[b]) *Voet, A.*: Chem. Rev. *20*, 169 (1937)
[c]) *Morris, D. F. C.*: J. Inorg. Nucl. Chem. *6*, 295 (1958), from lattice energies and lyotropic numbers, recalculated for the standard state considered
[d]) *Payzant et al.*: Can. J. Chem. *49*, 3309 (1971), from experimental electron affinities
[e]) Authors' estimates.

explanation of these observations is that a large hydration heat, ΔH_{hyd}, means a considerable ordering of water and is therefore opposed by a considerable entropy change, ΔS_{hyd}, of ion solvation. There is then a rough linear relationship between ΔH_{hyd} and $T\Delta S_{hyd}$ which is such that the overall ΔG_{hyd} of an anion (or a cation) follows ΔH_{hyd}. This free energy of hydration is a major term preventing the binding of anions (compare cations) by biological polymers, for when an anion enters a protein binding to a positive centre it must lose hydration, at least in part. The binding process is favoured by the entropy gain of the freed water (the hydration of the ion was an unfavourable entropic step) but opposed by the loss of ΔH_{hyd}. For heavily hydrated anions (small and highly charged), there will be considerable compensation of the loss of ΔH_{hyd} by the gain in $T\Delta S$, but as $\Delta H_{\text{hyd}} > T\Delta S_{\text{hyd}}$ a considerable interaction with the protein, usually through charged centres and H-bonding, is also required, although the overall ΔH may remain unfavourable. For a large anion, $T\Delta S$ for uptake may be small or even unfavourable and now *overall favourable* ΔH of binding is essential, requiring greater *relative* contribution of ΔH to binding to the protein than was the case for small anions. This can only be achieved by interaction with a *buried* positive charge which is not compensated inside a protein for these anions do not form good H-bonds. We shall see these factors below when we have described hydration in greater detail. The free energy of the binding of anions to proteins can now be broken up, firstly into an electrostatic term, which is important for binding of all anions to positive charges in proteins. The hydration energy (treated in electrostatic terms) will be opposed by the electrostatic interaction in the protein for there a neutralisation of charge occurs: $P^+ + L^- \rightleftharpoons PL$ where P is the protein and L the anion. In addition to this term there will be a second term, involving specific binding of the anion to the protein, which can be due to covalent bonding to a metal ion (of a metallo-protein), or to selective hydrogen-bonding to a non-metal cation of an amino-acid side chain such as $-NH_3^+$. Finally, the third term, which is also specific, is a steric restriction due to the repulsion or misfitting of the anion to the available positively charged hole in the protein. The free energies which control the binding of an anion to a protein can then be written in terms of the free energy of the binding reaction $P^+ + L^- \rightleftharpoons PL$:

$$\Delta G_{\text{binding}} = \Delta G_{\text{elec}} + \Delta G_{\text{specific}}$$

where each ΔG term refers to the difference in free energy between the hydrate and the bound protein site.

In the absence of $\Delta G_{\text{specific}}$ we can write an electrostatic free energy difference treating the water and the protein as media of different dielectric constants.

$$\Delta G(\text{binding}) = \Delta G(\text{protein site}) - \Delta G(\text{hydration})$$

$$= z \left\{ \frac{A}{\epsilon_1 r_1} - \frac{B}{\epsilon_2 r_2} + \frac{C}{\epsilon_2 r_3} \right\} = \Delta G_{\text{elec}}$$

where z = charge on the anion, ϵ_1 = dielectric of the protein (organic) solvent and r_1 is the distance between the anion and the positive site(s) in the protein, ϵ_2 is the dielectric constant of water, and r_2 the distance between the anion and the centre of the water dipole; A and B represent constants – the charge at the protein site and the dipole moment of water. $C/\epsilon_2 r_3$ is the energy of residual hydration of the positive charge in the protein site. It is known that the hydration energy of anions is quite well treated quantitatively by such a simple electrostatic picture where the free energy of hydration is proportional to the ratio charge/effective-radius (*23a*); anions having a very high value of this ratio e.g. F^-, CO_3^{2-}, PO_4^{3-} obviously belong in one group while I^-, Br^-, NO_3^-, ClO_4^- belong in another, that of the lowly hydrated anions. There are also anions of intermediate character, such as $CH_3CO_2^-$.

This type of equation is very similar to that used in the discussion of the binding of anions of different size to various cations of a fixed size in the formation of insoluble salts or of complex ions from their free aqueous ions in water. Two orders are found in such cases: (1) Given a large poorly hydrated cation, e.g. Cs^+, NEt_4^+, then it is precipitated or it forms complexes preferentially with very large anions, e.g. ClO_4^-. (2) Given a small cation, e.g. Li^+, then it forms precipitates or complexes with small anions, e.g. F^- and OH^- (*23b*). In accord with expectation based upon a purely electrostatic model of hydration and metal-ligand binding the large anion/large cation complexes are driven by overall favourable ΔH while small anion/small cation complex formation is driven largely by overall changes of $T\Delta S$. With respect to their charge/size ratio some of the cation sites of proteins, such as (hydrophobic) imidazole$^+$, resemble Cs^+, while other cationic sites in proteins, such as (hydrated) Fe^{3+}, resembles Li^+, and so different anion binding orders are to be expected in proteins.

Before discussing the selectivity of these binding sites in more detail we turn to a description of observed binding of anions to proteins to see if such different binding orders are in fact observed for different proteins. Anion binding to proteins has been frequently studied experimentally since the time when Hofmeister observed the "lyotropic" series in an examination of the effectiveness of anions upon the swelling of many proteins (*24, 25*). This series, of decreasing swelling produced, is

$$ClO_4^- > \begin{matrix} CNS^- \\ I^- \end{matrix} > \begin{matrix} ClO_3^- \\ N_3^- \end{matrix} > \begin{matrix} NO_3^- \\ CNO^- \end{matrix} > Br^- > Cl^- > \begin{matrix} CH_3CO_2^- \\ F^-, OH^- \end{matrix} > \begin{matrix} SO_4^{2-} \\ HPO_4^{2-} \end{matrix}$$

Perhaps the best quantitative binding study of this series is that of Fridovich who examined anion binding (inhibition) to the protein (enzyme) acetoacetic decarboxylase from C. *acetibutylicum* (*26*). Here the anions bind to a positive site of $pK_a \cong 6.0$ which is thought to be a histidine. Histidine sites are known to be prominent in the binding of anions to many proteins, e.g. lysozyme, ribonuclease, and cytochrome *c*, from crystallographic studies, see below. The binding constants, to acetocetic decarboxylase, $\log K$, for a series of anions are given in Table 20, and follow closely the Hofmeister series.

The study of the thermodynamics of this anion binding by Fridovitch and others has revealed that the main contribution to the binding of an anion at the top of the series, e.g. ClO_4^-, SCN^-, is the enthalpy change in the reaction, and it is opposed by the entropy change. Binding of anions toward the other end of the series, e.g. IO_3^-, is opposed by the enthalpy change, but is favoured (just) by the entropy change $T\Delta S$. Thus the relative values of ΔS and ΔH uptake are just as predicted from the simple view of electrostatic binding to a large (i.e. poorly hydrated) cation. Highly hydrated anions (large $-\Delta H$ (hydration)) cannot be taken up by such a protein centre as their

Table 20. Thermodynamic constants of inhibition of acetoacetic decarboxylase by various anions at *pH* 5.2

Ion	$\log K$[a]	ΔG[a] cal. mole^{-1}	ΔH cal. mole^{-1}	ΔS e.u.
F^-	1.04	$-1,410$	$-2,800$	-4.4
Cl^-	1.49	$-2,020$	$-11,850$	-33.0
Br^-	2.15	$-2,900$	$-17,500$	-49
I^-	3.35	$-4,550$	$-18,550$	-47
NO_3^-	3.23	$-4,390$	$-17,200$	-43
ClO_3^-	2.59	$-3,490$	$-16,300$	-43
ClO_4^-	3.45	$-4,690$	$-17,500$	-43
SCN^-	4.39	$-5,950$	$-28,000$	-74
BrO_3^-	1.27	$-1,730$	$-5,600$	-13
IO_3^-	0.99	$-1,350$	$+1,030$	$+8$
Cl_3CCOO^-	1.41	$-1,920$	$+3,440$	$+18$
Φ_4B^-	3.16	$-4,290$	$+7,130$	$+38$

Source: *Fridovich, I.:* J. Biol. Chem. *238*, 592 (1963)
[a]) $\log K$ and ΔG values were calculated from ΔH and ΔS data for $T = 25\,^\circ C$

uptake is opposed by too large an unfavourable ΔH due to the release of their *bound* water molecules. In fact, the lower the dielectric constant of the region of the protein which binds the anions, i.e. the more hydrophobic the site, the less well do the highly hydrated ions bind. The Hofmeister series is therefore usually thought of as being generated by hydrophobic binding. The series is then the reverse of the hydration free energy, e.g.

$$I^- > Br^- > Cl^- > F^- ;$$

$$ClO_3^- > BrO_3^- > IO_3^- > CO_3^{2-} \gg HPO_4^{2-}$$

Using hydration free energy as a guide line, other anions expected (and found) to bind well to such sites are CNO^-, N_3^-, etc. (Selectivity of binding amongst these anions can be based on steric constraints too, see below.) Amongst those expected and found to bind poorly are OH^-, $ROPO_3^{2-}$, $P_2O_7^{4-}$ and others.

(Before going further, it is worth observing that a site which gives the Hofmeister series through loss of hydration of an anion need not *necessarily* be *hydrophobic*. It need only be situated such that there is a prevention of *an anion which has a strong hydration* from entering the particular site, while weakly hydrated anions are bound.)

The anions which do not bind at the above types of sites are the highly hydrated anions such as F^-, SO_4^{2-}, and HPO_4^{2-}. Now, in certain enzymes and proteins, just these anions bind (indeed they are substrates), and there is no competitive binding of I^-, ClO_4^- etc. so that the reverse of the lyotropic series is found. For example, in most

Table 21. Protein-anion interactions

Protein	Ion in native structure	Ligands	Other Ions binding to this site
Carbonic anhydrase	OH^-	Zn^{2+}, Th 197	I^-, Br^-, Cl^-
	?	Lys 22, Arg 243	I^-, Br^-, SCN^-, RSO_2NH
α-Chymotrypsin	SO_4^{2-}	Ser 195, Tyr 146, His 57	SeO_4^{2-}
	SO_4^{2-}	Asn 95, Lys 177	SeO_4^{2-}
	SO_4^{2-}	Arg 154	SeO_4^{2-}
	SO_4^{2-}	NH_3^+, Ala 149	SeO_4^{2-}
	SO_4^{2-}	Asn 245	SeO_4^{2-}
	SO_4^{2-}	Asn 236	SeO_4^{2-}
Elastase ($pH = 5$)	SO_4^{2-}	Ser 195, His 57, N' 193	$2 H_2O$ ($pH = 8.5$)
	SO_4^{2-}	Arg 145, Arg 230	empty ($pH = 8.5$)
Lactate dehydrogenase apoenzyme	SO_4^{2-}	Arg 173	citrate
Myoglobin	SO_4^{2-}	His E7	
Ribonuclease	SO_4^{2-}	His 12, His 119, N' 120, H_2O	PO_4^{3-}, AsO_4^{3-}
	SO_4^{2-}	Lys 7, Lys 41	

From *Liljas, A., Rossman, M.:* Ann. Rev. Biochem *43*, 475 (1974)

kinases, dehydrogenases and haemoglobin, SO_4^{2-} and pyrophosphate bind strongly; the binding is often highly specific and all the evidence points to basic groups such as arginine and lysine (and only to a limited extent histidine) being present at the sites, Table 21. It is well known from protein structure determinations that arginine and lysine, which form extensive hydrogen-bonds, usually sit in extremely polar, heavily aquated, regions of an enzyme, but that histidine is found more frequently in non-polar regions. There are therefore two types of general positively-charged anion sites — non-polar (often histidine) and polar (often lysine or arginine) which usually give the lyotropic series and its inverse respectively. We must suppose, however, that the deciding factor is not the type of amino acid which carries the charge but *the region* of the protein in which the positive charge *sits* which decides whether the order of binding is that of the lyotropic series or quite the reverse.

Now, although it can be seen why a positive charge in a hydrophobic pocket should have a high affinity for anions (in the order of the Hofmeister series) through the electrostatic potential, (the dielectric constant is very low in the protein, making it desirable to secure charge neutralisation, see above equation), it is not so clear why a *hydrophilic* site should bind anions at all, although the relative oder of binding different anions can be reasonably surmised. The reaction

$$A^+ + L^- \rightleftharpoons AL$$

where A^+ is a simple amino-acid side-chain such as that of an arginine or lysine in free aqueous solution, e.g. a simple primary amine, is associated with only a very low affinity for all anions even such as sulphate and phosphate. Thus a single hydrophilic site in a protein may not be able to bind anions. Turning to this type of binding site in proteins, as in Table 21, it is seen that the sites often contain *several* positive charges with hydrogen-bonding potential and thus the binding involves the *chelation* through hydrogen bonds of several $-NH_3^+$ groups to one anion. In this way it is apparent that sufficient positive charge density can be built up together with high H-bond potential in a hydrophilic environment. Additionally, by a suitable arrangement of the groups, high specificity of anion binding can be achieved, e.g. for sulphate SO_4^{2-} rather than phosphate HPO_4^- or RCO_2^-, and ATP could be favoured over say ADP or AMP. The interaction under consideration is still overwhelmingly electrostatic although there can now be additionally exclusion, $\Delta G_{specific}$, on the basis of shape and size. *Chelation* of an ion by a protein, like the uptake of ions by a ring chelate, can have a great selectivity due to mismatching of the size of the ion and the 'hole'; this problem does not usually arise significantly when the binding is by a monodentate ligand.

Reverting to the electrostatic model for anion binding we predict that the binding of these anions with high ratio of charge/radius will be driven by overall favourable $T\Delta S$ terms. Unfortunately no thermodynamic data are available on this point.

Anion binding can also be assisted by specific recognition of an organic moiety, e.g. adenosine joined to say a phosphate group as in AMP, adenosine monophosphate, but here chelation is to the protein surface and one arm of the chelate is the anion while the other is the adenine. This is the same type of specificity that was seen for cobalt in vitamin B_{12}. *Specificity* of protein binding of anions (themselves) which are *low* in the Hofmeister series must arise from chelation (through H-bonds) and steric constraints. It is, then, largely a property of the stereo-chemical arrangements of arginines and lysines and to a lesser degree of histidines in the protein which dominates selectivity and of course this is a property of the amino acid sequence of the protein, see diphosphoglycerate binding to haemoglobin (*27*) and pyrophosphate binding to alcoholdehydrogenase (*28*). In the case of anions at the other end of the lyotropic series the steric constraints are brought about by bulky alkyl or similar side chains of very different amino-acids.

Basic amino-acid side-chains provide just one particular type of positive charge and we turn now to the case of metal ions, cations, which provide quite different positive centres. Clearly, they present centres which on first inspection have a much higher charge density (positive) and would be expected to bind in the opposite sense to that of the lyotropic series. As far as their electrostatic terms are concerned we need only note the value of the positive charge on the metal plus ligands and, to some extent, the size of its radius. This is controlled by the nature of the metal itself, of course, but also by the oxidation state of individual metals, (compare Fe(II) and Fe(III)) and by the protein ligands of the metal which can be anions. Table 22 gives a list of known charges (and net charges) on metal ions in enzymes. For the most highly charged centres the electrostatic term favours binding of small anions, F^- for example, but there is now one additional factor to be considered, namely covalent binding.

Table 22. Binding sites in proteins and net charge of the site

Protein	Metal	Site residues	Binding[a] Charac-teristics	Charge in metal ion	Net charge of site
Carp Albumin	Ca	Four carboxylates	S, f	$+2$	-2
Thermolysin	Ca	Three carboxylates	M, f	$+2$	-1
Bacterial nuclease	Ca	Three carboxylates	S, f	$+2$	-1
Insulin	Ca	Carboxylates of different chains	W, f	$+2$	0
Lysozyme	Ca	Two carboxylates	W, f	$+2$	0
Concanavalin	Ca	Two carboxylates	M, f	$+2$	0
Concanavalin	Mn	Histidine and two carboxylates	M, f	$+2$	0
Conalbumin	Fe	Nitrogen and three(?)tyrosine	S, s	$+3$	(0)
Ferredoxin	Fe	Two cysteine and sulphides	S, s	$+2, +3$	$-1, 0$
Rubredoxin	Fe	Four cysteine	S, s	$+2, +3$	$-2, -1$
Hemerythrin	Fe	Histidine, carboxylate, tyrosine	S, s	$+2$	?
Heme proteins	Fe	Four nitrogen of coenzyme and various nitrogen or sulphur ligands	S, s	$+2, +3$	$(0), (+1)$
Carboxypeptidase	Zn	Two histidine, one carboxylate	S, s	$+2$	$+1$
Insulin	Zn	Three histidine in different chains	M, f	$+2$	$+2$
Carbonic anhydrase	Zn	Three histidines	S, s	$+2$	$+2$
Alcohol dehydrogenase	Zn	Four cysteine	S, s	$+2$	-2
		Two cysteine and histidine	S, s	$+2$	0
Cu-albumin	Cu	Histidine and other N-ligands	S, s	$+2$	$(+2)$
B_{12} enzymes	Co	Five nitrogens of coenzyme and one carbon	S, s	$+3$	$+1$

[a]) S-strong, M-medium, W-weak, f-fast, s-slow

From: *Williams, R. J. P.*: Recent results in cancer research, Vol. 48. Berlin–Heidelberg–New York: Springer, 1974

4. Covalence

Covalency, as shown in the binding of ligands by metal ions, can be illustrated by data from the binding of free aqueous anions to free aqueous cations. There are two different orders; for class 'a' cations: $F^- > Cl^- > Br^- > I^-$, and for class 'b' metal ions: $I^- > Br^- > Cl^- > F^-$, respectively. Thus class 'b' metal ions bind in the sequence of the lyotropic series, roughly. The thermodynamic determinants of the orders are discussed in references (9, 10, 11) and here we note only that the order of binding to 'a'-class metal ion is largely due to overall $T\Delta S$ terms (different from the lyotropic series) and is usually overwhelmingly due to electrostatic terms, while binding to a 'b'-class metal ion is due to a particularly large favourable ΔH of binding (compare the lyotropic series), but this large ΔH is usually due to covalence and not electrostatic binding. (Note that the use of 'hard' and 'soft' acid/base concepts will lead to incorrect conclusions if a comparison of reasons for 'a' and 'b' classes is made with the reason for the lyotropic series.) Again we stress that this classification has strictly limited applicability because it is strongly solvent dependent and cannot be correlated with uniform specific characteristics of the several types of donors and acceptors (11) (see also below the case of Zn^{2+} in carbonic anhydrase). Now, the main cations concerned in biological systems are Mg(II), Ca(II), Cu(II), Cu(I), Zn(II), Co(II), Co(III), Fe(II), Fe(III), Mn(II), Mn(III), but only Cu(I) is a 'b'-class cation. Thus, for free aqueous ions, Cu(I) will amplify the lyotropic effect, while all other hydrated cations should oppose it to different degrees: Mg > Ca and Mg > Mn > Zn > Cu(II). We shall now consider metallo-enzymes in more detail, examining the evidence of binding of anions to the different metal cations of the proteins.

4.1. Metallo-Enzymes: Isolated (Single) Metal Ion Sites

Let us consider binding characteristics first. From an examination of the sites of metals in proteins given in Table 22 we can expect sites of the following kinds:
(a) Highly charged positive sites in water-free regions or regions of restricted access to water in a protein, e.g. carbonic anhydrase, where Zn^{2+} is bound to neutral ligands only but the metal site is deep in a narrow pocket.
(b) Charged positive sites in highly hydrated sites, e.g. carboxypeptidase, where Zn^{2+} is in open pocket.
(c) Lowly-charged sites as in alcohol dehydrogenase, where zinc is bound in a pocket to two anions RS^- (a similar case is that of Fe(II) in heme-proteins).
The last type of site will bind anions very poorly, as there is little electrostatic effect, and it is to be expected that binding agents for such sites will be neutral molecules which bind to metal ions, e.g. ortho phenanthroline and imidazole, or even those anions which form covalent bonds, e.g. RS^-, but not any of the usual anions of the Hofmeister series. This is observed in both heme-proteins and in alcohol dehydrogenase. Obviously this is not a good site for selective uptake or retention of anions.

113

The site (a), zinc in carbonic anhydrases, with an uncompensated metal ion in a deep pocket which can only be of low hydration, gives the series of binding of anions (29):

$$I^- > Br^- > Cl^- > F^- \; ; \; CNO^- > SCN^-, N_3^- \gg ClO_4^- > NO_3^- > \text{Acetate} \; .$$

i.e. close to the (lyotropic) series expected for a 'b' group cation, which zinc is not. Thus the protein pocket has changed the nature of zinc from an 'a'-class to a 'b-class cation. It is thought, in fact, that carbonic anhydrase can act as a good Cl^- transporting protein, even in the presence of either $SO_4^=$ and HPO_4^{2-}, for this very reason.

Curiously, with two particular anions carbonic anhydrase shows quite specific binding, showing very divergent behaviour from the above series. These anions are hydroxide and sulphonamides. (The two exceptions are removed if the metal in the enzyme, zinc, is replaced by copper (II), but is maintained if replaced by Co(II).) The two exceptional anions are the substrate (OH^-) of the enzyme and specific inhibitors which could be thought of as binding in a geometry which resembles the product of the enzyme reaction.

The binding of the sulphonamide group is aided by the presence of the organic moiety R, but this group can be varied very widely (30). The suggestion that the metal site is specially designed in some geometric fashion to accept the reactant (or an equivalent group of the right size and shape) means that anion uptake can be made highly, perhaps extremely, selective by making or putting a particular metal ion in a particular site.

The next case is that of a partially hydrated small metal ion in a metalloprotein which shows intermediate behaviour between 'a' and 'b' classes. In this class we find the single- or multi-copper centres of caeruloplasmin and laccase (31) as well as the zinc centre of carboxypeptidase (32). The binding order to these centres follows the order

Carboxypeptidase	$F^- \cong Cl^- \cong Br^-$
Laccase	$N_3^- \gg SCN^- > CNO^- \gg F^- \gg Cl^-$

Typical of metal (zinc) centres of this kind is that they bind simple carboxylate anions weakly, in accord with the fact that binding is very poorly selective on the basis of the

Hofmeister series, and that this binding can be greatly increased by increasing the hydrophobicity of the side-chain e.g. β-phenylpropionate as opposed to acetate (*33*) compare the binding of AMP to kinases, see above.

There is a further case of the binding of anions to proteins simultaneously with the uptake of a cation. The cations concerned must be those in fast exchange between the bound protein and the free aqueous state. The cations are then Mg^{2+}, Ca^{2+}, K^+, Na^+ and certain small organic cations. The most frequently met case is the binding of polyphosphates such as ATP and ADP (*34*). The sites are common to all kinases and some dehydrogenases have rather similar sites. The binding would appear to be to very hydrophilic protein regions containing additional positive charge due to basic side-chains of amino-acids. Thus the side-chains of the protein and metal cations cooperate in the binding of an anion, or, putting the case differently, the protein takes up an anion and a cation simultaneously. The sites are now very selective, for a multiple selectivity arises involving both anion and cation effects simultaneously.

The above discussion shows that highly selective anion uptake can be developed even though there are very few types of positive site which can bind anions. Now, once the anion enters the cell through a membrane it must be retained; as mentioned above, retention comes about through either redox reaction or/and condensation. One or two examples will be given in the following section.

5. Kinetic Control; Redox Uptake

The uptake of an anion followed by reaction is best described through the case of examples. Once Br^- and I^- have been taken into a cell, their size allows them to be selectively absorbed by enzymes much as in the uptake step. The fact that they have lower redox potentials than chloride also permits the reactions $Br^- \rightarrow Br^0$ and $I^- \rightarrow I^0$, e.g. by *plant* peroxidases. The redox potentials of these two anions are also very different in themselves. The reactive radicals so produced attack organic compounds readily, e.g. phenols, and Br and I, then appear in organic chemicals *covalently* linked. The enzymes responsible for this uptake are given in Table 23 and some of the compounds formed in Table 24.

Curiously iodine is actually liberated by such redox reactions in some plants as I_2, namely the sea-weeds, and it is said that this accounts for the smell of the sea-shore (not ozone); there is also the suggestion that this liberation of iodine is an

Table 23. Uptake of Halogens by Oxidation

Enzymes	Halide Oxidised
Horse Radish-Peroxidase	I^-
Lacto-Peroxidase	I^-, Br^-
Chloro- and Myelo-Peroxidase	I^-, Br^-, Cl^-

Table 24. Biological organic compounds containing halogens

Halogen	Protein containing halogen	Products of hydrolysis or metabolitis	Formula	Source
Chlorine	Not known	Geodin	$C_{17}H_{12}O_7Cl_2$	Aspergillus terreus
		Erdin	$C_{16}H_{10}O_7Cl_2$	Aspergillus terreus
		Griseofulvin	$C_{17}H_{17}O_6Cl$	Penicilium griseofulvum
		Caldariomycin	$C_5H_8O_2Cl_2$	Caldariomyces fumago
		Chlorotetracycline	$C_{22}H_{23}O_8N_2Cl$	Streptomyces aureofaciens
		Chloramphenicol	$C_{11}H_{12}O_5N_2Cl_2$	Streptomyces venezuelanus
Bromine	Some horny support tissue (sclero-proteins)	3-monobromo-L-tyrosine	$C_9H_{10}O_3NBr$	Gorgonacea and sponges
		3,5-dibromo-L-tyrosine	$C_9H_9O_3NBr_2$	
		6,6-dibromoindigo (Tyrian purple)	$C_{16}H_8O_2N_2Br$	Murex gasthropodes
Iodine	Thyro-globulin	3-monoiodotyrosine (MIT)	$C_9H_{10}O_3NI$	
	Thyral-bumin	3,5-diiodotyrosine (DIT)	$C_9H_9O_3NI_2$	Thyroid hormones of vertebrates
	Iodinated sclero-proteins	3,3'-diiodotyrosine (T_2)	$C_9H_9O_3NI_2$	some vertebrates some sponges and algae
		3,3'-5-triiodotyrosine (T_3)	$C_9H_8O_3NI_3$	
		3,3',5'-triiodotyrosine (T_3')	$C_9H_8O_3NI_3$	
		3,5,3,5-tetraiodotyrosine (Thyroxine-T_4)	$C_{15}H_{11}O_4NI_4$	
		Monoiodohistidine	$C_6H_8O_2N_3I$	
		3,3-diiodothyronine	$C_{15}H_{13}O_4NI_2$	
		3,5,5-triiodothyronine	$C_{15}H_{12}O_4NI_3$	

Source: *Roche, J., Fontaine, M., Leloup, J.:* Halides. In: Comparative Biochemistry, Vol. 5, p. 493 (eds. M. Florkin, H. Mason). London: Academic Press 1963.

early immune response by such plants. In fact, in certain protective cells, there are high concentrations of special *chloro*-peroxidases and their function could be to produce Cl_2 (Br_2 or I_2) so as to poison invading organisms.

Very many anions became incorporated into organic material after reduction rather than oxidation. Typical series of reactions are the reduction of nitrate to ammonia, followed by reaction with suitable organic functional groups:

$$NO_3^- \rightarrow NH_3 + R\text{–}CHO \rightarrow RCH{=}NH \rightarrow RCH_2\text{–}NH_3^+$$

and that of sulphate, through a series of steps, finishing up as thiols:

$$SO_4^{2-} \rightarrow \text{intermediates} \rightarrow R\text{–}SH$$

Other anions are incorporated by direct condensation, e.g.:

$$PO_4H^{2-} + HOR \rightarrow RPO_4^{2-} + H_2O$$

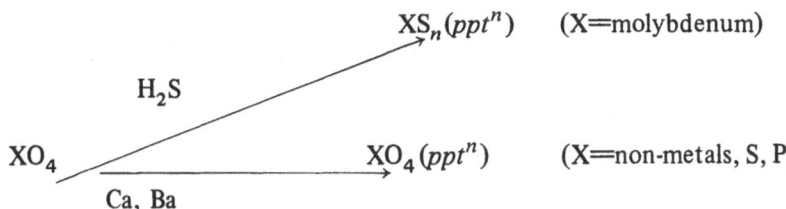

(note the occurrence of boron in boromycin) (*38*) and similar reactions can be written for silicates (note the ocurrence of silicon in mucopolysaccharides (*36*)).

Finally there are the cases in which the reduction can change an anion into a cation, introducing even greater opportunities for selective separation, or change an anion to a neutral form. Molybdenum and selenium uptakes allow a description of these cases.

On reduction of Mo (VI) to Mo (V) the latter shows a high affinity for both nitrogen and sulphur ligands and *Mitchell* and *Williams* pointed out that its chemistry begins to resemble that of Fe (III) (*37*). It would appear that when molybdenum is bound into proteins it is bound to sulphur and possible nitrogen ligands (*38*). In oxidation state VI, molybdenum readily reacts with H_2S to give sulphide precipitates; furthermore molybdenum even occurs as MoS_2 and related compounds. The separation of molybdenum from other anions is then closely related to the procedures of Analysis Tables, in which H_2S was used to reduce and precipitate molybdenate and to separate it from many other anions. A rough scheme of the possibilities is:

$$
\begin{array}{ccc}
 & & XS_n(ppt^n) \quad (X{=}\text{molybdenum}) \\
 & H_2S \nearrow & \\
XO_4 & \xrightarrow{} & XO_4(ppt^n) \quad (X{=}\text{non-metals, S, P}) \\
 & Ca, Ba &
\end{array}
$$

The above distinction would also separate V, W, As, Sb, (Bi), Se, and Ge, with Mo; and Si, (Al), (B), with S and P.

5.1. Uptake of Molybdenate

Molybdenate is in a very special category of anion. The element molybdenum, a metal, is available from oxygenated water as the anion MoO_4^{2-} and therefore is in competition with SO_4^{2-}, HPO_4^{2-} and possibly $H_2SiO_4^{2-}$. As an anion, molybdate is a weak acid which condenses with other groups forming for example the molybdo-phosphate polymer. Unlike sulphate and phosphate, Mo in these polymers sits in an octahedral hole, MoO_6, and it is possibly this change of coordination number which allows molybdenum to be separated from the other much more common anions. In fact, it is the only octahedral oxo-anion which can be made readily by biological systems.

Once inside a cell, molybdenum behaves very differently from the other anions (the nearest parallel must be arsenic, which is of course a poison but may play a role as a trace element). It is reduced in one electron steps certainly to Mo(V) but, maybe, under certain circumstances, reduced to Mo(IV) and Mo(III). Now, the oxidation states from (V) to (III) do not behave like the "anions" of non-metals. Every non-metal in anions which we have discussed keeps oxygen as its partner e.g.

$$SO_4^{2-} \rightarrow SO_3^{2-} \rightarrow S_2O_4^{2-} \rightarrow S$$

or it has a high affinity for hydrogen

$$NO_3^- \rightarrow NO_2^- \rightarrow N_2 \rightarrow NH_3$$

but there is little affinity between either N or S and other non-metals.

5.2. Uptake of Selenium

The redox diagram given earlier shows selenium to be an element which forms compounds which are readily reduced. It is similar to sulphur in this respect, but its reactions are kinetically much faster. Undoubtedly an uptake mechanism which takes in sulphate will take in selenium and then selenium will follow sulphur. It is known, however, that selenium is eventually segregated from sulphur, for it appears in an enzyme, glutathione reductase, as an essential component (39). The question is, how does this happen? The redox states of selenium offer a possible clue. In contrast to sulphur it has the oxidation states O to +6 which dominate its chemistry rather − 2 to + 6 like sulphur. (It is worth note that selenium could be an essential element merely because it undergoes two-electron redox reactions in the region of 0.0 volts and in which oxygen atom reactions can be involved, whereas no other element is capable of this same set of reactions at high rate. It is worth noting too that in organic chemistry SeO_2 is also a very special oxidising agent.) Thus selenium may be recognised and bound in an intermediate oxidation state, even in oxidation state zero (40).

118

VI. Conclusion

It is clear from the above that anions can be selectively absorbed and captured by biological systems as readily as cations. It is known that there are special proteins for the transport of anions such as chloride, sulphate, nitrate and phosphate across membranes. Given that there are such sites, the treatment of the thermodynamics and kinetics of competition follows that outlined for the cations. Overall, it is now possible to discern the methods by which biology absorbs, separates, and concentrates the twenty to thirty elements which it uses. The parallels between the chemistry and function of anions on the one hand and cations on the other is pulled together in Tables 25 and 26.

Table 25. Classification of cations in biological systems

	Na^+, K^+	Mg^{2+}, Ca^{2+}	Zn^{2+}	Fe, Cu, Co, Mo, V(?)
Major function	Charge carrier	Structure formers and triggers	Lewis-acid catalyst	Redox catalysts
Mobility	Highly mobile	Semi-mobile	Static	Static
Complexation	Weak complexes	Moderately strong complexes	Strong complexes	Strong complexes
Exchange rate	Very fast exchange	Moderately fast exchange	No exchange	No exchange
Ligand	Oxygen-atom binding	Oxygen-anion binding	Nitrogen/sulphur ligands	Nitrogen-sulphur ligands

Adapted from *Williams, R. J. P.:* Quarterley Reviews *24*, 339 (1970).

Table 26. Classification of anions in biological systems

	Cl^-, (NO_3^-)	CO_3^{2-}, F^-, SO_4^{2-} PO_4^{3-}, BO_3^{3-}, SO_4^{2-}	RO^-	RS^-, (SeO_3^{2-})
Function	Charge neutralisation	Structure formers	Acid catalysts	Redox catalysts
Mobility	Highly mobile	Semi-mobile to static	Static	Static
Complexation	Weak complexes (electrostatic)	Weak to moderately strong complexes (through H-bonds and condensation)	Strong complexes (covalently bound)	Strong complexes (covalently bound)
Exchange rate	Very fast exchange	Moderately fast exchange	No exchange	No exchange
Ligand	Hydrophobic centres; transition metals	Mg^{2+}, Ca^{2+}; Hydrogen bonding; ester formation; condensed structures	Covalent bonds to carbon; transition metals	Covalent bonds to carbon; transition metals

J. J. R. Fraústo da Silva and R. J. P. Williams

References:

General references:

1. *Bowen, H. J. M.:* Trace elements in biochemistry. London and New York: Academic Press 1966.
2. *Hewit, E. J., Smith, T. A.:* Plant mineral nutrition. London: The English Universities Press 1975.
3. *Underwood, E. J.:* Trace elements in human and animal nutrition, 3rd. ed. London and New York: Academic Press 1971.
4. Trace element metabolism in animals. Edinburgh and London: (eds. C. F. Mills, E. & S. Livingstone) 1970.
5. Trace element metabolism in animals – 2 (ed. W. Hoekstra, J. W. Suttie, H. E. Ganther, W. Merz). Baltimore, Maryland: University Park Press 1974.
6. Micronutrients in agriculture (eds. J. J. Mortvedt, P. M. Giordano, W. L. Lindsay). Madison, Wisconsin: Soil Science Society of U.S.A. 1972.
7. *Clarkson, D.:* Ion transport and cell structure in plants. London: McGraw Hill 1974.
8. *Diamond, J., Wright, E. M.:* Biological membranes: The physical basis of ion and non-electrolyte selectivity. Ann. Rev. Physiol. *31*, 581–646 (1969).
9. Inorganic biochemistry (ed. G. L. Eichhorn). Amsterdam: Elsevier, Sci. Pub. Co. 1973.
10. *Hughes; M. N.:* The inorganic chemistry of biological processes. New York: J. Wiley and Sons 1972.

Specific references:

1. *Thomson, A. J., Williams, R. J. P., Reslova, S.:* Structure and Bonding, Vol. *11*, p. 1. Berlin–Heidelberg–New York: Springer 1972.
2. *Leblanc, F., Robitaille, C., Rao, D. N.:* Quebec. Journal Hattori Bot. Lab. *38*, 405 (1974).
3. *Repaska, R.:* Biochem. Biophys. Acta. *30*, 225 (1958).
4. *Baddiley, J., Hancock, I. C., Sherwood, P. M. A.:* Nature *243*, 43 (1973).
5. *Nieboer, E., Ahmed, H. M., Puckett, K. J., Richardson, D. H. S.:* Lichenologist, *5*, 292 (1972).
6. *Neilands, J. B.:* Structure and Bonding, Vol. *1*, p. 59. Berlin–Heidelberg–New York: Springer 1966.
7. *Pardee, A. B.:* Science, *162*, 632 (1968).
8. *Mertz, W., Toepper, E. W., Roginski, E. E., Polanski, M. M.:* Fed. Proc. *33*, 2275 (1974).
9. *Ahrland, S., Chatt, J., Davies, N. R.:* Quart. Rev. (London) *12*, 265 (1958).
10. *Ahrland, S.:* Structure and Bonding, Vol. *5*, p. 118. Berlin–Heidelberg–New York: Springer 1968; Vol. *15*, p. 167 1973.
11. *Williams, R. J. P., Dale, J. D.:* Structure and Bonding, Vol. *1*, p. 249. Berlin–Heidelberg–New York: Springer 1966.
12. *Kruck, T. P. A., Sarkar. B.:* Inorg. Chem. *14*, 2382 (1975) and references therein.
13. *Vallee, B. L., Williams, R. J. P.:* Proc. Natl. Acad. Sci. (U.S.A.) *59*, 498 (1968).
14. *Fridovich, I.:* Ann. Rev. Biochem. *44*, 147 (1975).
15. *Banaszak, L. J., Watson, H. C., Kendrew, J. C.:* J. Mol. Biol. *12*, 130 (1965).
16. *Chen, K., Wang, J. H.:* Bioinorg. Chem. *3*, 339 (1974).
17. *Richardson, J. S., Thomas, K. A., Rubin, B. H., Richardson, D. L.:* Proc. Natl. Acad, Sci. U.S.A. *72*, 1349 (1975).
18. *Hall, D. O., Rao, K. K., Cammak, R.:* Sci. Prog. (Oxf.) *62*, 285 (1975).
19. *Malmström, B. G., Reinhammer, B., Vänngard, T.:* Biochem. Biophys. Acta. *205*, 48 (1970).
20. *Williams, R. J. P.:* Inorg. Chim. Acta Rev. *5*, 137 (1971).
21. *Schneider, W.:* Structure and Bonding, Vol. *23*, p. 123. Berlin–Heidelberg–New York: Springer 1975.
22. *Harrison, D. E., Taube, H.:* J. Am. Chem. Soc. *89*, 5706 (1967).
23. *Phillips, C. S. G., Williams, R. J. P.:* Inorganic Chemistry, Vol. *1* (a) p. 160, (b) p. 258. Oxford: 1965.
24. *Hofmeister, F.:* Arch. Exp. Pathol. Pharmacol. *24*, 247 (1888).

25. *Voet, A.:* Chem. Rev. *20*, 169 (1937).
26. *Fridovich, I.:* J. Biol. Chem. *238*, 592 (1963).
27. *Arnone, A.:* Nature *237*, 148 (1972).
28. *Branden, C.I., Eklund, H.E., Nordström, B., Boiwe, T., Söderlund, G., Zeppezauer, E., Ohlsson, I., Åkeson, Å.:* Proc. Natl. Acad. Sci. U.S.A. *70*, 2439 (1973).
29. *Pocker, Y., Stone, J. T.:* Biochemistry *6*, 668 (1967); *7*, 2936 (1968).
30. *Coleman, J. E.:* Carbonic Anhydrase. In: Inorganic Biochemistry, Vol. I. (ed. G. Eichhorn). Amsterdam: Elsevier Sci. Publ. 1973.
31. *Fee. J. A.:* Structure and Bonding, Vol. 23, p. 1, Berlin—Heidelberg—New York: Springer 1975 and references therein.
32. *Ludwig, M. L., Lipscomb, W. N.:* Carboxypeptidase A and Other Peptidases. In: Inorganic Biochemistry, Vol. I (ed. G. Eichhorn). Amsterdam: Elsevier Sci. Publ. 1973.
33. *Coleman, J. E., Vallee, B. L.:* Biochemistry *3*, 1874 (1964).
34. *Cohn, M., Hughes, T. R.:* J. Biol. Chem. *237*, 176 (1962).
35. *Dunitz, J. D., Hawley, D. M., Miklos, D., White, D. N. J., Berlin, Yu., Marusik, R., Prelog, V.:* Helv. Chim. Acta *54*, 1709 (1971).
36. *Shwarz, K.:* Proc. Natl. Acad. Sci. U.S.A. *70*, 1608 (1973).
37. *Williams, R. J. P., Mitchell, P. C. H.:* J. Chem. Soc. 1912 (1960).
38. *Bowden, F. L.:* The biochemical function of molybdenum. In: Techniques and topics in bioinorganic chemistry (ed. C. A. McAuliffe). London: The McMillan Press 1975.
39. *Rotruck, J. T., Pope, A. L., Ganther, H. E., Swanson, A. B., Hafeman, D. G., and Hoekstra, W. G.:* Science *179*, 588 (1972).
40. *Frost, D. V., Lish, P. M.:* Ann. Rev. Pharmac. 259 (1975).

121

The δ-Aminolevulinate Dehydratases: Molecular and Environmental Properties

Albert M. Cheh* and J. B. Neilands

Department of Biochemistry, University of California, Berkeley, California 94720, U.S.A.

Table of Contents

---

* Present address: Freshwater Biological Institute, University of Minnesota, Navarre, Minnesota 55392, U.S.A.

A. M. Cheh and J. B. Neilands

I. Introduction

The second enzyme in the heme biosynthetic pathway, δ-aminolevulinate (ALA)[1]) dehydratase (5-aminolevulinate hydrolyase, EC 4.2.1.24), has been under study for over twenty years. Recent findings, however, have sharply changed our knowledge of the properties of the mammalian liver enzyme. In addition, an environmental signifi-

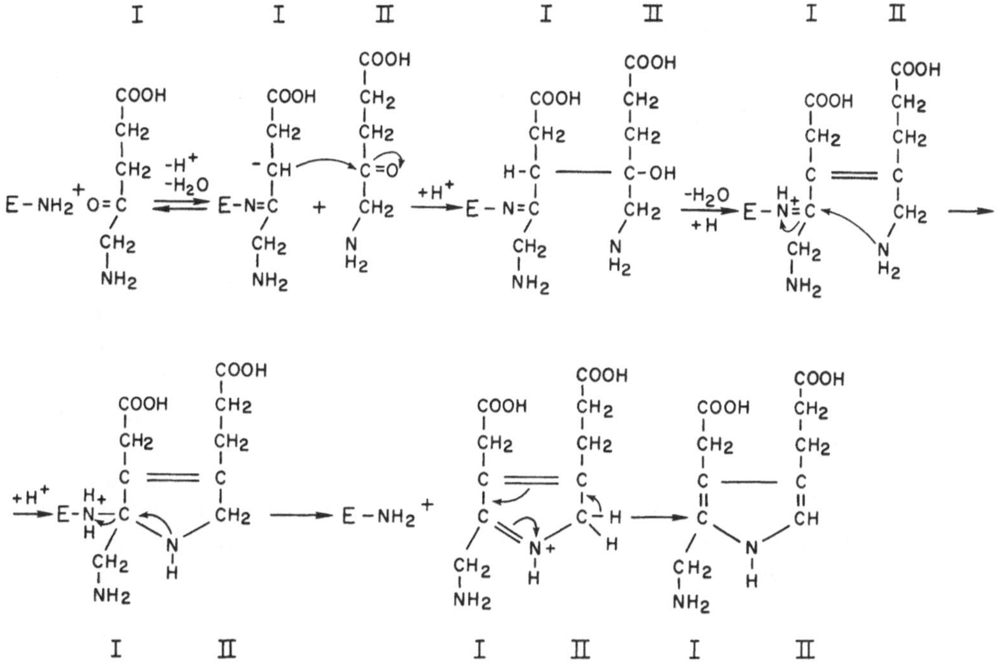

Fig. 1. Mechanism of ALA–D. From (2)

---

[1]) Abbreviations used:

ALA	δ-aminolevulinic acid	8–HQ	8-hydroxyquinoline
ALA–D	δ-aminolevulinic acid dehydratase	K_m	Michaelis constant
ATCase	aspartate transcarbamylase	LADH	liver alcohol dehydrogenase
BCS	bathocuproinedisulfonate	2–MET	2-mercaptoethanol
CD	circular dichroism	PBG	porphobilinogen
cySH	cysteine	SDS	sodium dodecylsulfate
E–L–M	enzyme-ligand-metal complex	SH	sulfhydryl
EDTA	ethylenediaminetetraacetic acid	SOD	superoxide dismutase
FDP	fructose-1,6-diphosphate	Tris	trishydroxymethylaminomethane
GSH	reduced glutathione		

125

cance of the human erythrocyte enzyme, namely its inhibition by Pb^{++}, has now been known for several years. It is the purpose of this review to describe recent findings pertaining to the biochemical and environmental properties of ALA dehydratase; emphasis will be placed on recent findings about the beef liver enzyme. For a survey of earlier work, reference is made to a review by *Shemin (1)*.

Fig. 1 illustrates the reaction catalyzed by ALA dehydratase (ALA–D) as proposed by *Nandi* and *Shemin (2)*. Two molecules of ALA are condensed, with the loss of two molecules of water, to form the product porphobilinogen (PBG). Three distinct reactions appear to be catalyzed by the same enzyme; described in order, these are: aldol condensation, loss of H_2O from the intersubstrate C–C-bond, and ring closure with loss of H_2O via intersubstrate Schiff base formation *(1)*.

II. Assay Methods

Two general methods have been used to assay ALA–D activity; both are dependent upon formation of a colored adduct upon addition of Ehrlich's reagent (p-dimethyl-aminobenzaldehyde) to PBG. Several workers *(3–9)* have employed one developed by *Gibson et al. (3)*. The Gibson assay contains phosphate buffer, pH 6.8, with a thiol activator, and ALA; $CuSO_4$ is used to stop the reaction. Following centrifugation, the supernatant is mixed with an equal volume of regular Ehrlich's reagent.

Most investigators *(2, 10–37)* have used variations of an assay first reported by *Granick* and *Mauzerall (10)*; it also contains phosphate buffer (Tris-HCl, pH 8 to pH 9 is used in the study of bacterial and plant enzymes, which have higher pH optima), thiol, and ALA, but uses a trichloroacetic acid- $HgCl_2$ stop solution. The supernatant after centrifugation is mixed with a modified perchlorate Ehrlich's reagent *(11)*. A mixed assay following *Gibson et al.* through the addition of $CuSO_4$ and then color development with the modified perchlorate Ehrlich's reagent, has also been reported *(38)*.

The general preference for the *Mauzerall* and *Granick (11)* modified assay is due to their report of the following drawbacks to the use of the regular Ehrlich's reagent: 1. the curve of color development vs. PBG concentration departs from linearity at $A_{552nm} > 0.2$, thereby limiting the useful range of the assay, 2. regular Ehrlich's reagent ($\epsilon_{552nm} = 3.4 \times 10^4$ M^{-1} cm^{-1}) is less sensitive than the modified reagent ($\epsilon_{555nm} = 6.2 \times 10^4$ M^{-1} cm^{-1}), 3. the color produced with the regular reagent is less stable than that produced by the modified reagent, and 4. Hg is more effective than Cu in tying up SH groups which would otherwise interfere with color development.

However, it was found *(9)* that the deviation from linearity reported by *Mauzerall* and *Granick (11)* was due to use of 6N HCl instead of 5N HCl. In this case, color development reaches a maximum very rapidly; at 5 minutes after mixing, when they took their reading, color had faded substantially, resulting in nonlinearity. The 5N HCl regular Ehrlich's assay, with *peak* readings taken, was linear up to an A_{552nm} of 0.7.

Comparisons of duplicate assays, with absorbances from 0.1 to 0.6, showed essentially no difference between the regular or modified Ehrlich's assay. In the usual modified assay, adding the stop solution results in a 1 to 2 dilution; that offsets the higher extinction coefficient of the modified reagent. Finally, there never has been any problem finding the peak color development with regular Ehrlich's reagent nor does it seem that copper is less effective in tying up SH groups under routine conditions.

Two features commend the Gibson assay: 1. The modified reagent is unstable. Usually, it is made up fresh each time before use, although *Granick et al.* (*39*) now state that the modified reagent is stable at room temperature for at least several weeks. 2. There is no mercury handling and disposal problem.

We have routinely used an adapted Gibson method as follows: Enzyme is incubated in 1.5 ml of 0.1M potassium phosphate pH 6.8 + 20mM 2–MET for 30 minutes at 37 °C in 12 ml conical glass centrifuge tubes. Then 50 μl of 0.1M ALA–HCl is added and mixed; the incubation is continued for 10–90 minutes. Two drops of saturated $CuSO_4$ are added and mixed to stop the reaction. Following centrifugation, the supernatant is added to 1.5 ml of 2% (w/v) p-dimethylaminobenzaldehyde in 5N HCl (Ehrlich's reagent) and mixed. The peak value of the A_{552nm} vs. time is converted to mμ-moles PBG using $\epsilon = 3.46 \times 10^4$ M^{-1} cm^{-1} (*3*). The assay is done in air; N_2 flushing or evacuation are not necessary.

There are, however, instances such as limited sample size or activity, where the higher extinction coefficient of the modified reagent is valuable, or where it is impossible to closely monitor peak color production with the regular reagent; here, the Mauzerall and Granick assay is clearly preferred. Note should be made of a potential pitfall in the method. The original paper by *Granick* and *Mauzerall* (*10*) calls for phosphate buffer, pH 6.8 without identifying the cation. Most workers have specified sodium phosphate or Tris-HCl (in the case of bacterial or plant enzymes); no problems should arise from use of those methods. Others (*18, 21, 27, 29, 34–36, 40*), however, have included potassium ion, which forms an insoluble perchlorate precipitate that might give an artifactually high A_{555nm} (*9, 37, 41*). *Finelli et al.* (*37*) have questioned the reported activation of the *Rhodopseudomonas spheroides* dehydratase by K^+. However, Fig. 7 in *Nandi et al.* (*17*) tends to rule out the possibility that activation is artifactual, and in any case, the data with other monovalent cations indicate a genuine activation by them.

Bonsignore et al. (*42*) developed an adapted Granick assay for the erythrocyte enzyme. Workers who have studied the effect of Pb^{++} on the erythrocyte enzyme have split between those choosing the *Bonsignore* assay (*42–48*) and those employing a phosphate buffer assay instead (*37, 49–59*). More recent reports have tended to favor the use of phosphate. *Hernberg et al.* (*51*) found that substitution of sodium phosphate for bicarbonate in the Bonsignore method resulted in a 20% increase in activity and greater accuracy in the assay. *Finelli et al.* (*37*) reported that the bicarbonate used by *Bonsignore et al.* (*42*) is an inadequate buffer; they also recommend the use of sodium phosphate in its place. Again, as noted above, the use of potassium ion should be avoided.

127

The *Granicks et al.* (*39, 41*) describe a micromethod for assaying erythrocyte ALA—D and its inhibition by Pb^{++}; these papers are recommended for their detailed discussions of proper handling and controls to avoid artifacts. *Kneip et al.* (*60*) describe artifacts affecting assay of Pb^{++} inhibition of erythrocyte ALA—D that might arise from sampling techniques. Adsorption of metal to glassware or contamination with adventitious metals (in the case of ALA—D, Zn^{++} might cause an artifactual increase in activity (see *37, 61*)) has long plagued those studying trace metals. Precautions necessary for trace metal work have been covered by *Thiers* (*62*).

III. Purification

Several purification methods have been described or referred to in the literature: liver (*3, 4, 9, 12, 15, 19, 21, 32, 33, 40, 63*), erythrocytes (*10, 14, 40*), other animal cells (*5, 35*), *Rhodopseudomonas* (*17, 36, 38*), other bacteria (*13, 23, 30, 31*), yeast (*26, 64*), other fungi (*7, 24, 34*), and plants (*16, 20, 21, 22, 25*). These will not be discussed here except to say that it appears that at least 1000 fold purification of a crude homogenate of mammalian liver ALA—D is required to achieve homogeneity. In our hands, one published method (*33*) does not appear to be completely adequate; the rather high initial and final specific activities reported, as compared to other purification methods (*9, 19, 32*) could be due to $KClO_4$.

IV. Molecular Properties of Liver ALA–D

A. Molecular Weight, Subunit Weight, and Subunit Number

ALA–D from mammalian liver was generally thought to have a molecular weight of 235,000–275,000 daltons (*1, 15, 28, 29, 32, 33*) (similar values were found for erythrocytes (*14*), plants (*25*) and *R. spheriodes* (*1, 17, 29*)). Concurrent reports of a subunit weight of 39,500–42,250 (*1, 28, 29, 33*), and the presence of six substrate binding Schiff base sites (*28, 33*) led to the consensus of a hexameric enzyme.

Wilson et al. (*32*), however, reported a significantly different subunit weight of 35.500, which suggested the possibility of 7 subunits. Recently, evidence has been accumulated which indicates that liver ALA–D is an octameric enzyme rather than a hexameric one. It appears that the enzyme has a molecular weight of 280,000–290,000 (*9, 63, 65*) and a subunit weight of 34,500–36,500 (*9, 32, 63, 65–67*). The conclusive proof of an octameric structure is demonstrated in electron micrographs published by *Wu et al.* (*63*) indicating a cubic octameric structure with dihedral (D_4) symmetry.

B. Half-site Reactivity

Unpublished data have been cited (*9, 63, 65, 68*) to indicate that perhaps only half of the Schiff base sites react with substrate. If so, then ALA–D appears to exhibit the phenomenon of half-site reactivity (*69, 70*), a subject which has been thoroughly reviewed by *Seydoux, Malhotra* and *Bernhard* (*71*), and *Lazdunski* (*72, 73*).

The latest data on ALA–D conflict directly with earlier reports (*1, 28, 33*) indicating that six or seven Schiff base sites (seven or eight based on 280,000 daltons) may be counted with ^{14}C–ALA + NaBH$_4$. However, those reports used a single concentration of ALA well in excess of K_m [2]) and an excess of NaBH$_4$. Excessive ALA + NaBH$_4$ concentrations might cause an overflow labelling of sites that normally do not bind ALA, or in the extreme case, nonspecific labelling of random amino groups throughout the enzyme molecule. The latter is seen in the reaction between excess fructose-1,6-diphosphate (FDP) and FDP aldolase (*74*).

In cases where excess labelling might be a problem, titrations of residual enzyme activity vs. number of covalently labelled sites should be done. ^{14}C–ALA + NaBH$_4$ at concentrations 1/100 to 1/1000 of those used in earlier experiments substantially inactivated ALA dehydratase; titration suggests that only 4 sites rather than 8 need be labelled to cause virtual elimination of enzyme activity (Fig. 2).

---

[2]) $K_m = 1.4 \times 10^{-4}$ M. ALA–D is a bisubstrate enzyme, one bound covalently through a Schiff base, the other not. Assuming significant differential binding affinities, K_m measures the affinity of the more loosely bound substrate, which is presumably the non-covalently bound one. Therefore, ALA in those experiments must be in even greater excess than might be suspected.

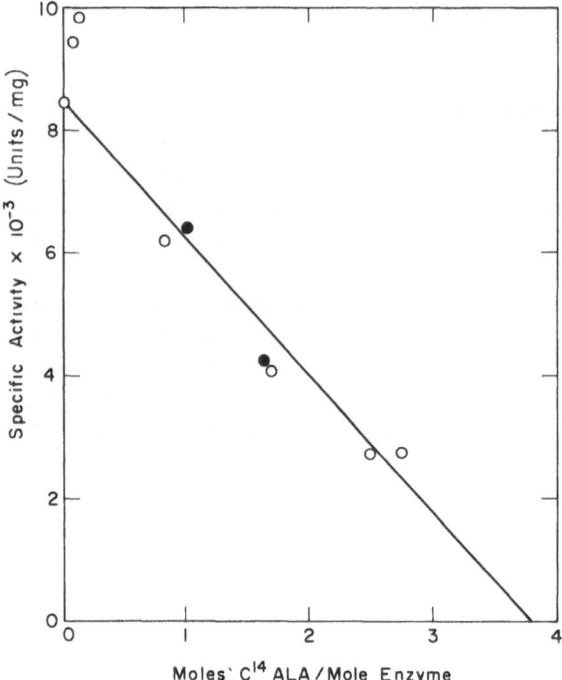

Fig. 2. Titration of Schiff base sites of beef liver ALA–D. The reaction conditions were 125 μg enzyme, 0.05 M potassium phosphate, pH 6.8, 7 mM 2–MET, varying amounts of ALA from 0 to 10^{-4} M, and 5×10^{-4} M NaBH$_4$ in 200 μl volumes. After 40 minutes at 25 °C, excess ALA and NaBH$_4$ were removed by (NH$_4$)$_2$SO$_4$ precipitation of the enzyme and dialysis.

● : EDTA present
○ : EDTA absent

Two points at the upper left are controls: ALA only, no NaBH$_4$ and no ALA, no NaBH$_4$.
From (9).

C. Molecular Basis for Apparent Half-site Reactivity

Four models[3]) that might give rise to different affinities for the two sites on a dimer were described by *Levitzki et al.* (70). In the case of ALA–D where the overall number of binding sites has not been established, another possibility must be considered, name-

---

[3]) The four models that give rise to different affinities for the two sites on a dimer may be described as:
(1) nonidentical subunits with different sites.
(2) asymmetric dimerization of identical monomers.
(3) steric hindrance of accessibility to closely placed sites (identical monomers).
(4) negative cooperativity mediated by ligand induced conformational change (identical monomers).

ly that there might be only four sites. Then, either two identical or nonidentical subunits give rise to an intersubunit site (perhaps each binding one of the two substrates) or ALA–D is $\alpha_4\beta_4$ where only the α subunits have active sites.

Only one subunit size was found by SDS gel electrophoresis (8, 9, 28, 33, 63, 65–67) or sedimentation equilibrium (32, 63). The number of tryptic peptides corresponds well with one + the number of arg + lys per subunit weight (28, 63). Therefore, if the nonidentical monomer model #1 of *Levitzki et al.* (70) is correct, the subunits must be quite similar with the differences thus far undetected.

Asymmetric dimerization is the explanation favored by *Seydoux et al.* (71) for the underlying mechanism that gives rise to half-site reactivity; it will have to await further evidence for either confirmation or refutation, in the case of ALA–D. Also, there is no evidence for or against steric hindrance being the reason for apparent half-site reactivity in ALA–D.

Asymmetric dimerization could give rise to an $\alpha_4\beta_4$ structure or an intersubunit site where nonidentical portions of the active site are created by two identical monomers. Therefore, the finding of 8 identical subunits does not rule out a four active site enzyme.

The fourth model of *Levitzki et al.* (70) (negative homotropic interactions), which is favored by them, requires substrate induced conformational changes which result in negative cooperativity in substrate binding. A change in the circular dichroism (CD) spectrum between 260 and 290 nm is seen upon addition of ALA to the dehydratase (9); it remains to be established if it is due to conformational changes. Interestingly, the change is complete by the time $[ALA] = K_m$, suggesting that CD might be measuring interaction between ALA and the Schiff base site.

The electron micrographs of ALA–D (63) are interesting in light of possible half-site reactivity; observation of single and intersecting bifurcations between subunits of the cubic oligomer suggests the presence of four dimeric units, assuming that the single bifurcation pattern is not due to incomplete staining (63). Fig. 3, an adaptation of the

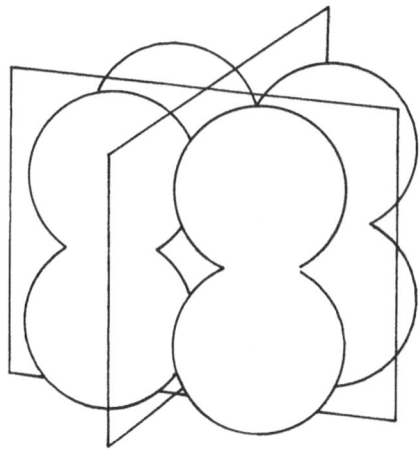

Fig. 3. A model of beef liver ALA–D. Adapted from (63).

131

model proposed by *Wu et al.* (*63*), illustrates a model for ALA–D that would give rise to the staining pattern observed. As noted above, either asymmetric dimerization (*71*) or negative cooperativity in binding to the two protomers of a dimer (*70*) could give rise to half-site reactivity. Such interactions could occur through a set of particularly extensive or tight binding domains, identifiable by lack of stain penetration. A four active site enzyme could also arise from interprotomer construction of sites in the same binding domains.

Since ALA–D has an oligomeric molecular weight of 280,000–290,000 daltons, a preliminary observation (*4*) of the presence of units of 140,000 and 70,000 daltons could indicate dissociation. Those data are suspect because 140,000 was reported to be the oligomeric weight. No detailed description of the data or confirmation of the observation has been reported. It remains to be seen if ALA–D can be broken down into dimers.

The available evidence does not establish whether there are four or eight active sites. Up to 8 sites may be titrated with substrate + $NaBH_4$, but it remains to be seen if the second set of four sites are actual active sites; equilibrium binding studies may give the answer. Given the widespread occurrence of half-site reactivity, it would not be surprising to find that ALA–D is in fact a half-site enzyme. The active site number has a bearing on the metal ion stoichiometry and enzyme mechanism, which are discussed below.

V. Catalytic Properties of Liver ALA–D

A few new findings in this area have been reported since the appearance of the review by *Shemin* (*1*).

As noted earlier, the enzyme binds two substrate molecules and K_m measurements may only reflect binding by a lower affinity site.

A. Sulfhydryl Requirements

It appears that every ALA–D, regardless of source, requires the presence of a thiol activator when assayed after the first steps of purification (*1*). Our findings on the effectiveness of various thiol activators (*9*) parallel those of *Wilson et al.* (*32*), except we found that 2-MET is an excellent activator. Iodoacetamide, p-hydroxymercuribenzoate, and N-ethylmaleimide have all been shown to be potent inhibitors of ALA–D (*1, 3, 4, 32*). It appears that there are essential sulfhydryl groups which are readily oxidizable and highly reactive. The role of added thiols is complicated; in the case of beef liver, preincubation is required, but with the enzyme from other sources, preincubation is not necessary. The identity of the thiol may also influence the behavior of added metal ions (see below).

B. Possible Role of a Histidine Residue in Catalysis

A sharp drop in activity of liver ALA–D is seen on the acid side of the optimum *pH* of 6.5–6.8 (*1, 3, 9, 32*). This could arise from metal removal or protonation of a catalytic residue such as histidine. *Tsukamoto et al.* (*67*) reported evidence for the presence of an essential histidine in liver ALA–D. Photooxidation with methylene blue destroyed two histidine residues, with concomitant loss in activity; other types of residues were unaffected. Titration of residual activity vs. histidines modified during inactivation by diethyl pyrocarbonate indicates that modification of only one highly reactive histidine residue suffices to eliminate enzymatic activity. Half maximum rates of photoinactivation are observed at *pH* 6.8. That *pH* is substantially higher than the pK_a (6.0 to 6.3) for half maximum enzyme activity; therefore, the relationship of the histidine residue to the *pH* profile is uncertain.

A mechanism proposed by *Nandi* and *Shemin* (*2*) for ALA–D from *R. spheroides* (Fig. 1) was described earlier. A similar mechanism may apply to the liver enzyme. Undoubtedly, a basic group participates in the reaction by removing a proton from C–3 of one ALA molecule, aiding in the production of the carbanion that undergoes aldol condensation with a second ALA. In rabbit muscle FDP aldolase, precisely such a role may be performed by a histidine residue (*75*); possible parallels between ALA–D and FDP aldolases are discussed below.

133

C. Activation by NH_4^+ and Divalent Metal Ions

We have noted that often a greater amount of enzyme activity seems to be present after ammonium sulfate precipitation steps in the purification (see *33*). This was traced to an activation by NH_4^+; the activation increment may be as great as 25% of the activity in the absence of NH_4^+ (*9*). This suggests a monovalent cation activation of the liver enzyme, although such an activation is not as dramatic as the one seen with the *R. spheroides* dehydratase.

Besides finding that Zn^{++} in the growth medium was required for the production of high levels of activity of *Ustilago sphaerogena* ALA–D (*6*), *Komai* also noted that the addition of Zn^{++} to the assay solution caused an increase in the activity of the partially purified enzyme (*76*). It is felt that the *in vitro* finding may be an artifact due to the use of cysteine or reduced glutathione (GSH) as the thiol activator. As seen in Table 1. zinc ion does not affect the apparent activity when 2-mercaptoethanol (2–MET) or Cleland's reagent are used as the thiol activators; there is an increase with Zn^{++} + GSH over GSH, and a large increase when Zn^{++} is added to a cysteine activated enzyme. It would appear that zinc ion brings activation by cysteine or GSH up to about the level of activation by 2–MET or Cleland's reagent. The fully purified enzyme also exhibits full activity (2–MET activation) without added Zn^{++}.

As noted by *Wilson et al.* (*32*), cadmium ion acts as an inconsistent activator.

Table 1. Effect of Zinc Ion on Beef Liver ALA–D
Activation by Different Thiols.[a])

Thiol	Aded Zinc	Activity
2–MET	none	137%
2–MET	10^{-4} M	137%
CySH	none	100%
CySH	10^{-4} M	149%
Cleland's Reagent	none	132%
Cleland's Reagent	10^{-4} M	136%
GSH	none	118%
GSH	10^{-4} M	131%

[a]) Partially purified enzyme was assayed in the usual
manner (see text). Activity with cysteine, no added
Zn^{++}, is defined as 100%. Taken from (*9*).

D. Metal Ion Inhibition

Even with those enzymes requiring metals, an excess of the essential metal is often inhibitory. This is true in the case of the dehydratase, not surprisingly, considering the essential SH groups present. Table 2. shows the inhibitory effect of excess Zn^{++} or

Table 2. Effect of Excess Zn^{++} and Co^{++} on Beef Liver ALA–D[a]

Concentration of Co^{++} in Assay	Activity	Concentration of Zn^{++} in Assay	Activity
none	100%		
2×10^{-6} M	100%		
5×10^{-6} M	95%		
10^{-5} M	85%	10^{-5}	100%
2×10^{-5} M	75%		
3×10^{-5} M	68%		
5×10^{-5} M	40%		
10^{-4} M	25%	10^{-4} M	90%
2×10^{-4} M	10%	10^{-3} M	25%

[a]) Purified enzyme assayed in the usual manner (see text). Taken from (9)

Co^{++}. Co^{++} acts as a noncompetitive inhibitor (9). The inhibition by Co^{++} could be related to the accumulation of ferrichrome and lowered ALA–D activity in *U. sphaerogena* grown in the presence of cobalt (77), and to the finding of decreased ALA–D activity in Co^{++} intoxicated *Neurospora crassa* (24). It might also be a reason for lowered catalase activity in liver after cobalt administration (78), although a recent paper suggests that Co^{++} may cause increased degradation of heme in the liver (79).

Other divalent ions known to bind strongly to sulfhydryl and other ligands are even more effective in inhibiting the dehydratase. Hg^{++} is used as an assay stop reagent. Cu^{++} is also lethal to the enzyme, besides binding strongly, it probably also catalyzes the oxidation of sulfhydryls. Most attention has centered on Pb^{++} due to its environmental and public health significance; this will be covered separately below.

VI. Metalloenzyme Nature of Liver ALA–D

A. Introduction – Criteria

Research on metalloenzymes and metalloproteins has accelerated in the last decade; studies on several zinc enzymes have reached a point where X-ray structures are known, and the exact position and function of the metal ligands and other catalytic groups are coming to be understood. First, however, one must establish that the enzyme in question is a *metalloenzyme*, i.e., that the metal ion is an integral and essential part of the enzyme.

This has generally been done by satisfying the following criteria used to define metalloenzymes (paraphrased and condensed from *Vallee (80)*, *Malmström* and *Neilands (81)*, *Mildvan (82)* and *Scrutton (83)*).

1. Native Metal

The native metal is found bound to the enzyme after purification. This is due to the extremely high affinity of the enzyme for its metal; affinity constants are typically 10^{10} M^{-1} or higher for zinc metalloenzymes (*84, 85*). By contrast, a *metal activated enzyme* forms a weaker complex, and often requires the presence of high concentrations of metal so that enough is bound to the enzyme to activate it fully; because of this, the metal is often lost upon purification. A typical affinity constant for zinc in the case of a metal activated enzyme is 10^5 M^{-1} (*86*).

2. Stoichiometry

Because of the tight binding of metal, the pure protein contains an integral stoichiometric amount of bound metal. Full activity is observed in the absence of further metal addition. Generally, during purification, the native metal content rises to the true stoichiometry, while extraneous metal contents fall as protein impurities are removed. Also, there is an integral ratio between metal ions and cofactors.

3. Essentiality

Chelating agents, which bind to the metal, and often remove it, cause inhibition (this is often the first indirect indication that an enzyme has a metal requirement). Removal of the essential metal results in a lowered activity which is directly proportional to the amount of metal still bound (e.g., *87*). A metal-free apoenzyme is inactive, and, by addition of metal, activity is regenerated in amounts directly proportional to the amounts of metal restored (e.g., *88, 89*).

In practice, the first two criteria are often not met. Perfect stoichiometry is rarely found, since there are often small losses during purification (e.g., *90*), and adventitious metals, including perhaps the native metal, may be picked up in the purification process. We will discuss the essentiality of metal ions to ALA−D, and the behavior of the enzyme regarding the above criteria.

B. Inhibition by Chelating Agents; Reversal of Inhibition by Metals

Gibson et al (*3*) gave the first indication of a metal requirement for beef liver ALA−D by reporting an inhibition by ethylenediaminetetraaceticacid (EDTA). However, 8-hydroxyquinoline (8−HQ) did not inhibit; inhibition by the sulfonic acid of 8−HQ, which was not fully reversed with Fe^{+3}, suggested to these workers that the sulfonic acid group was the actual inhibitor.

Komai and *Neilands* (*7*) reported inhibition of the beef liver enzyme by neocuproine and bathocuproinedisulfonate (BCS); the same inhibition pattern was found for the *U. sphaerogena* enzyme.

Subsequently, *Wilson et al* (*32*) reported the inhibition of the liver enzyme by EDTA, *o*-phenanthroline, and 2,2'-dipyridyl. *Gurba et al* reported inhibition by EDTA, 8−HQ and its sulfonic acid, 2,2'-dipyridyl, and *o*-phenanthroline (*33*); they also reported that *m*-phenanthroline, which is similar in structure to the *o*-isomer, but is not a chelator, is not an inhibitor. This is an important point because in the case of glutamate dehydrogenase, inhibition by *o*-phenanthroline, attributed to binding of Zn^{++} (*91*), was later shown to be a property of the aromatic structure (*92*).

Our findings with the beef liver enzyme are as follows (*9*): EDTA is by far the most effective inhibitor; by using acid washed glassware and metal-free buffers and reagents, complete inhibition can be observed with 10^{-6} M EDTA. The positions of dipyridyl, 8−HQ, *o*-phenanthroline, and BCS roughly parallel their affinities for metal ions. Chelators specific for Cu^{2+} compared to Cu^+ (cuprizone and sodium diethyldithiocarbamate) and Fe^{3+} compared to Fe^{2+} (rhodotorulic acid and deferriferrichrome A) have no effect at concentrations up to 5×10^{-3} M. *p*-Toluenesulfonate at 10^{-3} M also has absolutely no effect, in contrast to the suggestion of sulfonate inhibition made by *Gibson et al* (*3*).

Wilson et al (*32*) added metals at 10^{-4} M to liver ALA−D inhibited by 10^{-5} M EDTA. Of the metals Zn^{++}, Fe^{++}, Cu^{++}, Mg^{++}, and Ca^{++}, only Zn^{++} restored full activity. The amount of activity restored by the other metals was not described. Zinc at 10^{-4} M is borderline inhibitory to ALA−D; since similar levels of other metals might also be inhibitory, reconstitution of activity should be attempted with as low a concentration of metals as possible. Table 3. shows our findings (*9*) on the reversal of EDTA inhibition by various metals; it appears to establish the pattern: $Cd^{++} \geqslant Zn^{++} > Co^{++} > Ni^{++} > Mn^{++} > Fe^{++} \gg Fe^{+3}, Cr^{+3}, Mg^{+2} > Cu^{+2}$.

As noted by *Wilson et al* (*32*), Zn^{++} seems to be most effective in restoring full activity; however substantial reactivation is seen with other metal ions, so a possible

Table 3. Reversal of EDTA Inhibition of Beef Liver ALA–D by Addition of Metal Ions.[a]

Experiment I[b]		Experiment II[c]	
Added Metal Ion	% of Native Activity	Added Metal Ion	% of Native Activity
none	4%	none	6%
Cd^{++}	101%	Cd^{++}	96%
Zn^{++}	100%	Zn^{++}	82%
Co^{++}	90%	Co^{++}	78%
Ni^{++}	90%	Ni^{++}	51%
Mn^{++}	98%	Mn^{++}	35%
Fe^{++}	70%	Fe^{++}	15%
Fe^{+++}	11%		
Cr^{+++}	6%		
Mg^{++}	7%		
Cu^{++}	0%		

[a] Purified enzyme assayed in the usual manner (see text). From (9).
[b] EDTA present at 10^{-5} M. Metal ions added to a concentration of 2×10^{-5} M.
[c] EDTA present at 9×10^{-6} M. Metal ions added to a concentration of 10^{-5} M.

role for them cannot be ruled out on the basis of these data. The observation of inhibition by a diverse set of chelators, lack of effect of m-phenanthroline, and reversal of inhibition by metal ions does suggest that inhibition is due to metal chelation; still, more direct evidence of native metal identity, stoichiometry and essentiality would be desirable.

C. Metal Analysis – Stoichiometry

ALA–D was reported to have 1.1 g-atoms of Zn, traces (0.2–0.4) of Fe, and no detectable Co per mole of enzyme as determined by atomic absorption spectrophotometry on material prepared by Gurba et al. (33). Cu and other metals besides Fe and Co were not examined. Cu, but no other transition metals were found by Iodice et al. using only a partially purified enzyme (93), but they later reported Cu could be removed without loss of activity (94).

No consistent set of results for Zn, Cu, and Fe was ever obtained with our purified enzyme (9). Anywhere from 0.5 to 1.8 g-atoms of Zn/280,000 daltons, 0.1 to 1.0 of Cu, and 0.1 to 0.7 of Fe were measured in many different preparations. Occasional preparations showed much higher levels of all three, suggesting contamination. In the cases of liver alcohol dehydrogenase (LADH), finding 3.1–4.3 g-atoms of Zn (95), and yeast aldolase, finding 1.2–1.6 g-atoms of Zn (90), suggested stoichiometries of 4.0 and 2.0 respectively; here three different metals might be involved, the ranges are

Table 4a and 4b. Metal Content of Purified Preparations of Beef Liver ALA–D[a])

a) Preparations Using Ammonium Sulfate to Concentrate and Store Column
Eluates.

Metal	Range of Findings (gm-atoms/mole of enzyme)
(many preparations:)	
Zn	0.5–1.8, up to 6.0 (avg. ~ 1.1)
Cu	0.1–1.0, up to 3.0 (avg. ~ 0.4)
Fe	0.1–0.7, up to 2.0 (avg. ~ 0.3)
(single preprations:)	
Mg	< 0.001
Mn	≤ 0.005
Ni	≤ 0.1
Co	≤ 0.05
Cr	≤ 0.03
V	≤ 0.1

b) Preparations Using Diaflo PM–10 Membrane Ultrafiltration to Concentrate
Column Eluates.

Metal	Preparation I (gm-atoms/mole of enzyme)	Preparation II (gm-atoms/mole of enzyme)
Zn	1.6	2.4
Cu	0.25	0.35
Fe	0.2	0.5
Cd	≤ 0.1	–
Mn	–	≤ 0.01
Ni	–	≤ 0.07
Co	–	≤ 0.05

[a]) Purified enzyme was dialyzed against metal-free 0.05 M tris-HCl pH 7.2. Metal
contents were determined by atomic absorption spectroscopy. From (9).

larger, and a stoichiometry of 1.0 seems unreasonable for an octameric enzyme.
Table 4a summarizes the range of metal content data we (9) obtained by atomic
absorption spectroscopy analysis of ALA–D produced in the usual manner, with
$(NH_4)_2 SO_4$ used to precipitate and store column eluates.

D. Apoenzyme

Despite the range of 1.2–1.6 g-atoms of zinc observed in purified yeast aldolase, a
stoichiometry of 2.0 seems justifiable because an inactive apoenzyme can be titrated
back to full activity with 2 g-atoms of Zn. Most of the zinc could be removed from
ALA–D by passage through Chelex, by dialysis against chelators and then metal free

139

buffer, or by treatment with chelators and passage through a Biogel P—6 column, (90) 0.9 × 15 cm, equilibrated with 0.05 M metal free Tris—HCl pH 7.2. EDTA was chosen for production of apoenzyme both because very low amounts are needed to remove metal and, unlike the aromatic chelators, it is not likely to attach to the enzyme via hydrophobic bonding.

According to Simpson and Vallee, *Escherichia coli* alkaline phosphatase was believed to contain 4 Zn^{++} and require 4 metal ions (Zn or Co) for complete reconstitution of apoenzyme (89). However, *Applebury* and *Coleman* reported that 2 was the probable stoichiometry (96). *Csopak* and *Szajn* found a possible explanation for the discrepancy (97, 98). EDTA was used in the preparations of Simpson and Vallee; it was found to bind strongly to alkaline phosphatase and was not removed by dialysis. Bound EDTA could be responsible for the higher stoichiometry reported by *Simpson and Vallee*.

Binding of EDTA to superoxide dismutase and failure to remove it by ordinary dialysis against acetate buffer was also observed (88); EDTA binding to proteins may be a commonplace phenomenon (99). To rule out the possibility that EDTA might bind in the COO^- and $^+NH_3$ substrate sites in ALA—D, experiments were done to demonstrate that the apo protein did not contain bound EDTA. Through the use of 2-Acetate ^{14}—C—EDTA, we found (9) that essentially all of the EDTA was removed from the apoenzyme by either dialysis or Biogel P—6 chromatography.

E. Reconstitution

In addition to the unusually low and variable metal ion stoichiometry in the native enzyme, another puzzling phenomenon was repeatedly observed. Despite removal of 90% of the metal found in the purified protein, and the inhibition by chelators of many diverse structures and types, the apoenzyme appeared to be fully active in an assay mixture whose components were specially treated to remove metals. Adventitious contamination is probably responsible. *Kobes et al.* also noted the impossibility of controlling contamination at the 10^{-8} M level. They assayed apoaldolase without activators, and with a temperature reduction to 15 °C; this allowed the use of much higher enzyme concentrations in the assay (90). Assaying ALA—D at 0 °C allows the use of $10^{-6} M$ instead of $10^{-8} M$ enzyme; this overcomes the contamination problem. In Fig. 4 it is seen that the addition of up to 5 and 6 g-atoms of zinc titrates activity of now inactive apo-dehydratase to a maximum value. The titration is more or less linear, with activity reflecting metal content. This evidence satisfied the necessary criteria for showing that ALA—D requires metal ions (probably Zn^{++}) for activity. A "native" enzyme with about 1.1 zinc and 0.7 (Cu + Fe) per mole shows less than 1/4 of the activity obtained by titration with added zinc. It would appear that purified "native" enzyme is at least 75% apoenzyme.

Other metals were tried in the low temperature assay at ratios of 4 and 6 metals/ enzyme. Only Cd^{++} gave activity identical to Zn^{++}. Co^{++} and Mn^{++} restored up to 25%

140

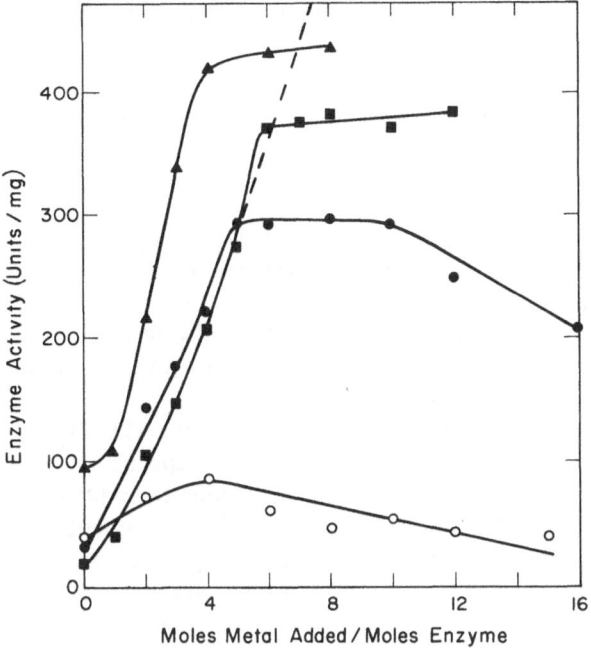

Fig. 4. Titration of purified and apo–ALA–D (beef liver) with Zn^{++} and Co^{++}.

▲: Purified "native" ALA–D (preparation I) assayed at 10^{-6} M in 0.05 M sodium phosphate, pH 6.8, with Zn^{++} added.

■: Apo–ALA–D (preparation II) assayed at 2×10^{-6} M in 0.05 M Tris-HCl, pH 7.2, with Zn^{++} added.

●: Apo–ALA–D (preparation III) assayed at 1.5×10^{-6} M in 0.05 M Tris–HCl, pH 7.2, with Zn^{++} added.

○: Apo–ALA–D (preparation III) assayed at 1.5×10^{-6} M in 0.05 M Tris–HCl, pH 7.2, with Co^{++} added.

The dotted line shows the number of Zn^{++} ions that would be required for reconstitution if the maximum activity of native ALA–D is higher than the activity plateaus observed. All assays contained 20mM 2-MET. Adapted from (9).

of the activity. Ni^{++} showed 0–10% reactivation. Cu^+ as the chloride, Cu^{++}, Fe^{++} and Fe^{+++}, Cr^{+++}, Mg^{++}, and Ca^{++} were ineffective in restoring any activity.

The activity obtained with Cd^{++} suggests the metal binding site accommodates different sized ions. It would be very surprising if only d^{10} ions were active, however; one would expect Co^{++} (which is so similar to Zn^{++} that they are frequently interchangeable (100, 101)), and possibly Mn^{++} to be as active as Zn^{++}. Excess metal is inhibitory, and it appears that the inhibition by Co^{++} sets in sooner than that by Zn^{++}. Carboxypeptidase (85) and alkaline phosphatase (84) have a 10^3 lower affinity for Co^{++} than for Zn^{++}; ALA–D may exhibit similar behavior. This, combined with the fact that Co^{++} starts to become inhibitory at 10^{-5} M, comparable to the amounts used

141

in the low temperature assay, would mean that full reconstitution could not occur without a superimposed inhibition.

F. Discussion of Metal Ion Stoichiometry

Caution is required in interpreting the stoichiometry data. There is no "native" enzyme to compare with the reconstituted one: it is assumed that reconstitution regenerates a true native enzyme. It is clear, however, that the active sites number either 4 or 8, arising from 8 probably identical subunits; the observation that 5–6 Zn^{++} reconstitution leads to maximum activity is not in accord with the other molecular stoichiometries.

This disagreement may be resolved in either of two ways. 1) Only four Zn^{++} are required per octameric enzyme; 5–6 must be added to the apoenzyme because kinetic or competitive (with substrate, or thiol, etc.) factors require the presence of excess Zn^{++} for full reconstitution. 2) Eight Zn^{++} are required per octameric enzyme. Some denaturation and loss of enzyme activity may occur during purification and dialysis (or purification, metal removal and dialysis to remove chelators, in the case of an apoenzyme). The presence of inactive enzyme would result in the regeneration of apparent maximal activity with the addition of *less* than the correct stoichiometric amount of metal. In fact, this is seen during titration of different apoenzyme preparations. Preparations that reach a lower maximum activity plateau require less added Zn^{++} to reach that plateau. It is entirely possible that an observed stoichiometry of 5–6 might reflect an actual stoichiometry of 8 (see dotted line in Fig. 4). The difficulty arises from the lack of native material which might indicate what the true maximum activity plateau should be. It should be noted in Fig. 4, that the apoenzyme curves cannot be extrapolated to the activity seen with "native" enzyme, because the assays used different enzyme preparations and were done under different conditions; native enzyme was assayed at a pH closer to the optimum, and would be expected to yield a higher specific activity even if it contained an equal amount of inactive material.

We favor the latter explanation, which suggests a stoichiometry of eight. Clearly, however, the question is still unsettled, but is of critical importance because of its bearing on the reasons for apparent half-site reactivity (see above) and the role of metal in ALA dehydratase (see below).

G. Spectra of Cd^{++} ALA–D

Reconstitution of apoenzyme with cadmium results in some interesting changes in the absorption and CD spectra. Based on studies of metallothionein and model complexes, *Kägi* and *Vallee* (102) claimed that the finding of an absorption increase at 250 nm upon replacement of zinc by cadmium is characteristic of cadmium mercaptides.

142

Cd^{++}–LADH contains a 245 nm absorbance which was suggested to have arisen from a cadmium mercaptide chromophore charge transfer (*103*). It is now known from X-ray studies that LADH contains two cysteines and one histidine about the two catalytic zinc atoms and four cysteines about the other two zinc atoms (*104*). In aspartate transcarbamylase (ATCase), the finding of a charge transfer absorption centered at 250 nm, supported by the finding of differential reactivity of apo and metalloenzymes to SH reagents, has led to a claim of SH binding of the metal (*105*). Cd^{++} spectra were not reported, but such a differential reactivity is also seen with aldolase (*106*), an enzyme that may have structural similarities with ALA–D.

Replacement of Zn^{++} with Cd^{++} in ALA–D results in the appearance of an absorption band centered around 235 nm (Fig. 5a). Appearance of a positive CD band below 260 nm, whose maximum is impossible to pinpoint since it is superimposed upon a rapidly decreasing curve, indicates an asymmetry about the metal binding site (Fig. 5b); the metal in ALA–D may have a distorted ligand geometry. It would be desirable to show a differential reactivity of SH groups in comparing apo and metalloenzymes; until such supporting evidence is generated, one may only speculate about the ligands bound to zinc in ALA–D.

H. ALA–D as a Zinc Metalloenzyme

It appears that liver ALA–D might be a zinc metalloenzyme although it is still unsettled as to whether the stoichiometry is 4 or 8. It would be tempting but unjustifiable to claim that the traces of iron and copper found in the purified protein represent adventitious contamination. Since these traces might be all of a large *in vivo* content that remained after purification, one cannot rule out a possible role for those metals. Inhibition of the *Ustilago* enzyme with BCS, a reagent specific for Cu$^+$ (as opposed to Cu^{++}) suggested to *Komai* and *Neilands* (*7*) that the former ion may be required, although the authors stressed that further work was required to establish this point.

If any functional role for Cu^{n+} or Fe^{n+} should exist, however, it would be very difficult to prove with the enzyme as presently isolated; Cu$^+$ and Cu^{++} are inhibitory at the $10^{-7}M$ to $10^{-8}M$ level (probably binding to and/or oxidizing essential sulfhydryl groups), making reconstitution practically impossible. Only the isolation of a native copper protein could allow one to demonstrate a functional role for Cu. Similar problems probably exist with attempted reconstitution with Fe^{++} or Fe^{+++}. At this time, all that may be said is that while 1) the enzyme as isolated possesses more Zn^{++} than Fe^{n+} or Cu^{n+}, 2) loss of Zn^{++} may be due to the purification method (see below), 3) we can propose reaction mechanisms involving Zn^{++} that are very similar to those Zn^{++} is known to participate in other metalloenzymes (see below), 4) there seems to be no need for Fe or Cu, metals which are more useful when the mechanism involves electron transfer, and, 5) *in vivo* studies (see below) argue against a role for Cu, we have no proof that beef liver ALA–D has (only) Zn^{++} as its native metal.

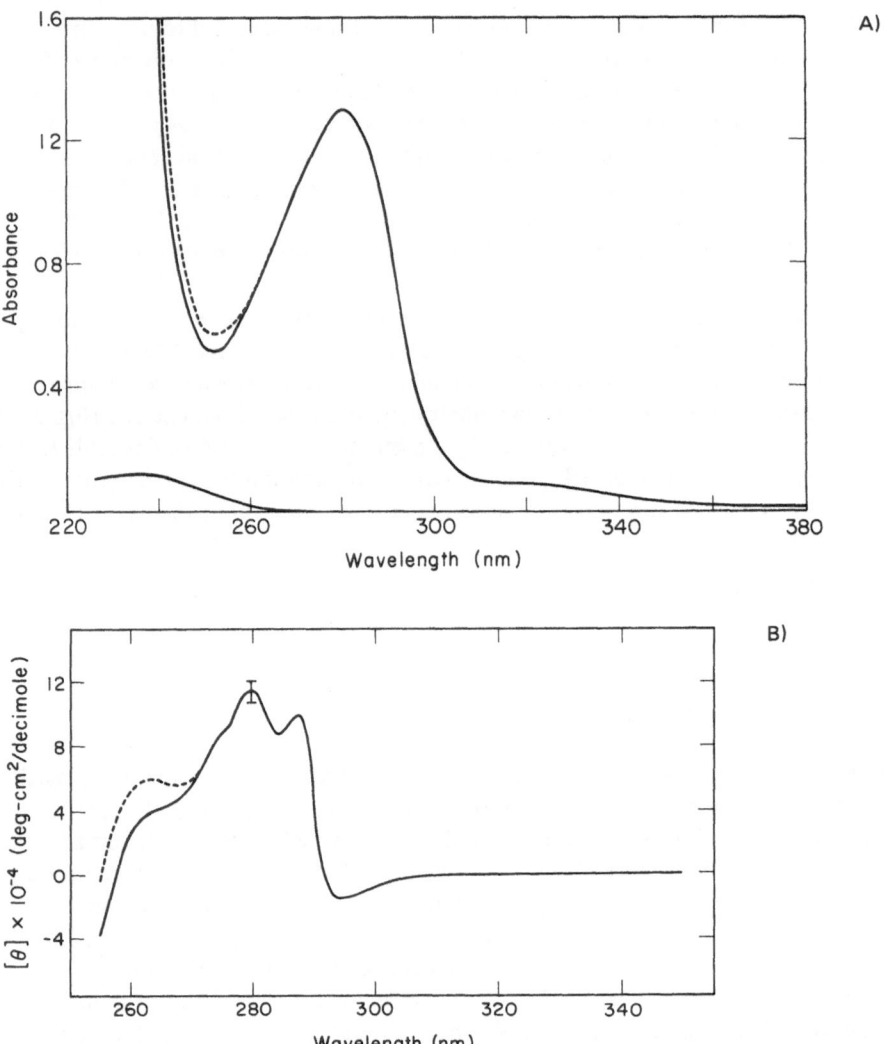

Fig. 5. Spectra of Cd^{++} ALA–D (beef liver).
A) Absorption spectra of ALA–D – 1.1 mg/ml in 0.05 M potassium phosphate, pH 6.8. Upper solid line – apo–ALA–D or enzyme reconstituted with 4 Zn^{++}/mole, dotted line – change upon reconstitution with 4 Cd^{++}/mole, lower solid line – difference spectrum between 4 Zn^{++} and 4 Cd^{++} ALA–D.
B) CD spectrum of ALA–D. Solid line – apo–ALA–D or 5–6 Zn^{++}/mole ALA–D, dotted line – replacement of Zn^{++} by Cd^{++} (5–6/mole). From (9).

144

I. Discussion of the Metalloenzyme Defining Criteria

Definitions of *metallo* and *metal activated enzymes* were originally written to distinguish between the two types. At the time, the distinction between metalloenzymes ($K_{affinity} = 10^{10}\ M^{-1}$) and metal activated ones ($K_{affinity} = 10^5\ M^{-1}$) were obvious. It was easy to make clear and rigid definitions for each. Because there might be enzymes with intermediate affinities (80), *Mildvan* (82) has suggested a dividing line of 10^7 to $10^8\ M^{-1}$. One might ask why one should be concerned with drawing such a distinction; obviously, there is little use for the ability to make a purely semantic division calling some metal requiring enzymes *metallo* and others *metal activated*. The value of defining a metalloenzyme lies in the use of the definition in a predictive manner. If one identified an enzyme as being a metalloenzyme, one could predict that the metal would be a tightly bound integral part of the enzyme which could be studied directly in its native state in the purified enzyme.

Vallee (107) states that "the metal atoms of metalloproteins are *bound so firmly that they are not removed from the protein by the isolation procedure*" (italics his). In practice, because of the greater value of a predictive description, the Vallee definitions have been applied as follows: when tight binding is observed, indicating that a particular enzyme is a metalloenzyme, then one predicts the native metal will be found bound to the enzyme in the correct stoichiometry after purification. Retention of bound metals during dialysis has generally been the criterion used to identify metalloenzymes. The identity and stoichiometry of the bound metals are determined; these data are then believed to reflect the native state of the enzyme.

Apo beef liver ALA–D was fully active in the standard assay, despite enzyme and metal both being present at $< 10^{-7}\ M$, and dialysis failed to remove bound metals from the purified enzyme. This suggested a very large affinity for the native metal, particularly since 20 mM 2–MET was present in the assay, so beef liver ALA–D apparently was a metalloenzyme which should be in the "native" state after purification. Possible loss of up to 25% of the metal during purification, such as occurs with aldolase, was considered; however, the loss of over 75% of the native metal seemed unlikely.

From the usual manner in which the metalloenzyme defining criteria are used, it would seem that the approximately one Zn^{++} found bound to ALA–D after purification and dialysis might reflect the native metal ion stoichiometry. The reconstitution data, which indicate a minimum requirement for four Zn^{++} and a likely requirement for eight Zn^{++} rule out that interpretation. In retrospect, it appears that most of the metal *was* lost during purification, and not surprisingly since a relatively lengthy purification procedure was followed. At least some of the loss might be due to the use of ammonium sulfate which is present at $> 2M$ concentrations in several steps. In comparing the two parts of Table 4, one can see that substitution of Diaflo concentration for $(NH_4)_2SO_4$ precipitation of column eluates tends to raise the Zn^{++} content while lowering the Cu^{n+} and Fe^{n+} contents. Other purification procedures and the presence of 1 mM 2–MET in some of them may have also contributed to losses.

If such a common and routine procedure as ammonium sulfate precipitation removes metal ions, then that would be disturbing indeed. Isolation of purified metalloenzymes with less than stoichiometric amounts of metal might be a common experience. This has occurred with aldolase (90) and LADH (95) in addition to ALA–D. Yet one would be misled into believing that because of observed tight metal binding the less than complete stoichiometry of the purified enzyme is the correct one.

Care must be exercized when establishing the identity of the native metal and the stoichiometry of a metalloenzyme. It is common practice to purify a metalloprotein and report the identity and number of metals bound to it. It must be stressed that until supporting evidence is developed, no proof exists that these necessarily have anything to do with the function of the protein or are the correct identity and stoichiometry. We cannot emphasize too strongly a point made by *Scrutton*: "Valid information cannot be obtained from single determinations of the metal content of enzymes at *any* level of purification . . . " (83) (italics ours). Because of the importance of developing such supporting evidence, perhaps one should not make the prediction that metalloenzymes retain their native metals upon purification; that obviously depends on the drastic nature of the purification steps.

Scrutton (83) discusses some other problems that may arise if one follows the metalloenzyme criteria uncritically. For example, the *sum* of the two metals, e.g., $Co^{++} + Zn^{++}$ in yeast alcohol dehydrogenase (108), or $Mn^{++} + Mg^{++}$ in chicken liver pyruvate carboxylase (109) may equal integral stoichiometry under certain unusual conditions of nutrition even though integral stoichiometry is not observed for either member of the pair. Also, the content of the native metal may not rise during purification if the contaminants removed have even higher contents of the same metal.

One might question the justification for dividing metal requiring enzymes into metallo and metal activated, if one must still confirm the native metal identity and stoichiometry to assure that the enzyme is in fact in the "native state". There is a reason for making the distinction, but it is an *in vivo* one rather than an *in vitro* one. Does the level of available metal ions *in vivo* ever drop to the point where the enzyme exists partially as an apoenzyme, and enzyme activity falls? If so, then regulation of the available metal concentration could result in regulating enzyme activity. If the enzymic affinity for metal is so great that the enzyme is never an apoenzyme under *in vivo* conditions, then metal availability does not regulate enzyme activity. In this respect, a distinction between metallo and metal activated enzymes might also be valuable; *in vitro* studies however, could only suggest behavior *in vivo*.

Malmström and *Rosenberg* (110) have very correctly called attention to the fact that the functional role of metals in catalysis is probably the same in both metallo and metal activated enzymes and an overemphasis of differences based on metal-enzyme affinities is likely to obscure mechanistic or chemical similarities. That remains the heart of any study on metals in metalloenzymes — the elucidation of the role of the metal in influencing enzyme activity.

VII. The Role of Zinc in ALA—D

A. Introduction

Mildvan (*82*) and *Scrutton* (*83*) have classified metal ion roles on the basis of the nature of the coordination scheme of metal, ligands (including substrate) and enzyme. Physical methods may be used to distinguish among 1) $E–L–M$, 2 a) $E–M–L$, 2b) $E{<}^M_L$ and 3) $M–E–L$ coordination schemes and thus narrow down the possible roles of the metal ion, since certain coordinations suggest certain roles while excluding others. For example, 3) implies a structural role, but no direct participation of the metal in catalysis. With bisubstrate reactions, such as that catalyzed by ALA—D, the following schemes must be considered:

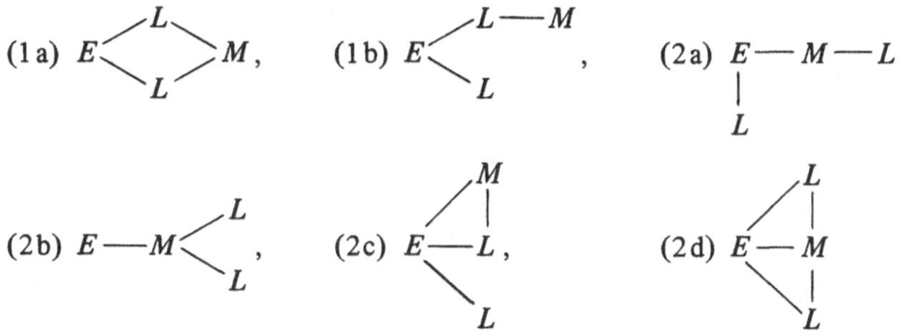

$$(1a)\ E{<}^L_L{>}M, \qquad (1b)\ E{<}^{L-M}_L, \qquad (2a)\ E—M—L$$

$$(2b)\ E—M{<}^L_L, \qquad (2c)\ E—L{<}^M_L, \qquad (2d)\ E—M{<}^L_L$$

and

$$(3)\quad M—E{<}^L_L$$

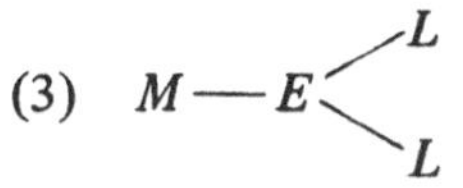

We already know somethin about the interaction of zinc and ALA—D (strong affinity for each other, thus excluding (1 a, b) and enzyme and substrate (probable binding of substrate through COO^- and NH_3^+, thus excluding (2a) and (2b). The choices are most likely limited to (2c), (2d), and (3); physical probing may help identify the coordination scheme. Since those experiments have not been done, we will discuss instead possible roles for Zn^{++} in ALA—D from a functional standpoint. The reader is referred to a recent review by *Dunn* (*111*) for a detailed discussion of the mechanisms of zinc catalyzed reactions.

   Metals in metalloenzymes may have one or more of the following roles: 1) *Structural* – stabilizing a particular enzyme conformation or subunit aggregation which is usually, but not always, required for catalytic activity. LADH is an example where two zincs are structural and two are catalytic (*112*). In ATCase, all the zincs are structural (*113, 114*). Zn^{++} is proposed to be an allosteric inhibitor of glutamate dehydrogenase (*115*); it could be an allosteric activator for other enzymes. 2) *Substrate binding* –

metal could aid in anchoring the substrate in the active site, with or without any particular catalytic function. Metal removal means substrate is no longer bound as well or fails to be oriented properly for reaction (template effect, (*111*)), with resulting loss in catalytic activity. Another possible substrate binding role is charge neutralization, allowing negatively charged reactants to overcome mutual repulsion (*111*). 3) *Catalytic-* metal participates directly in and is essential for the catalytic reaction.

B. Possible Roles for Metals in ALA–D

1. Structural

If ALA–D is a 4 zinc enzyme with 8 active sites, then the zincs might be structural; there would not be enough to act catalytically, unless two sites share a zinc ion. The problem of demonstrating a structural role becomes the detection of conformational differences between metallo and apoenzyme.

ALA–D from *R. spheroides* is K^+ activated, where K^+ changes the enzyme kinetics from a sigmoid to a hyperbolic v vs. s plot; K^+ involvement in changing the protein conformation is also suggested by incomplete labelling of the active site by ALA + $NaBH_4$ in the absence of K^+ (2). K^+ apparently functions as an allosteric effector.

In the beef liver enzyme, Zn^{++} is absolutely required for activity, rather than being involved in changing the nature of the enzyme kinetics; in any case a role for Zn^{++} as an allosteric effector should also be discernible through comparison of the conformations of apo and metalloenzymes. None have been detected thus far in comparisons between apoenzyme and Zn^{++} reconstituted enzyme.

2. Substrate Binding

In ALA–D, zinc ion could also have a substrate binding role as is suggested by the formation of a five-membered ring with substrate (Fig. 6a and 6b); or, it could help bind substrate carboxyl groups to the enzyme. In either case, one would expect that the apoenzyme would have a lowered affinity for ALA–D compared to the metalloenzyme. There is no evidence so far for or against a substrate binding role.

The enzyme binds equally well to an ALA type of affinity column in the presence of EDTA, and $NaBH_4$ reduction of ALA onto the metalloenzyme proceeds equally well in EDTA. Also, the CD spectrum shift obtained by adding ALA to the enzyme is the same with apo or reconstituted enzyme, so it seems metal is not required for at least some substrate binding. Affinity column and CD measured binding probably reflect binding in the Schiff base site rather than the metal site (Fig. 6a, b), however, since covalent Schiff base attachment predicts stronger binding in that site. Equilibrium dialysis may ultimately demonstrate or disprove a substrate binding role.

3. Catalytic

In most zinc metalloenzymes, such as the amide and ester hydrolases, the dehydrogenases, and metallo-FDP aldolase, there are catalytic zincs present (83, 107, 111). The metal in several other lyases is believed to act catalytically also (82, 83). Since ALA–D is an aldolase and a hydro-lyase, we think speculation on possible catalytic roles is justified. In fact, we think the analogy to other Zn enzymes is so strong that despite being unable to comment upon a structural or substrate binding role, we favor a direct catalytic role for Zn in ALA–D.

It has been noted (33) that beef liver ALA–D might be mechanistically like class I or II aldolases because of involvement of a Schiff base intermediate and a metal ion in catalysis, and that the metal might participate in dehydration. At the time those proposals were put forward, however, it was impossible to state how one zinc atom could accommodate six or seven active sites while remaining bound to the enzyme through two or more ligands; no detailed mechanisms were proposed.

Now that the metal ion stoichiometry has been revised, and it appears there are enough metal ions to accommodate all the active sites (through sharing between dimers if the stoichiometry is 4 instead of 8), one may write out some detailed enzyme mechanisms which demonstrate precisely the possible action of Zn^{++} in the various intermediate reactions and show how it is capable of performing all the roles previously suggested.

Mechanisms for carboxypeptidase (111), LADH (111), and aldolase (83) can be written whereby zinc acts as a Lewis acid polarizing a carbonyl function. An obvious place to put the zinc in ALA–D is shown in Fig. 6a, an expansion of the mechanism proposed by *Nandi* and *Shemin* (2) (Fig. 1). It is tempting to draw a five-membered chelate ring although it is not required in the mechanism, and in fact may not exist. Zinc could aid the aldol condensation *and* dehydration reaction by acting as an electron sink. After dehydration, the chelate ring is broken, and ring closure may then occur.

The simultaneous presence of the two functional groups, the Schiff base and zinc ion, however, allows the postulation of a second possible mechanism;[4]) the alternate mechanism is described in Fig. 6b. In it, zinc acts in a manner like that proposed for type II aldolases (116, 117) which contain zinc ions but no Schiff base intermediates. Zinc stabilizes a carbanion which now attacks a Schiff base bound substrate; therefore, zinc catalysis involves enhancement of substrate nucleophilicity (towards a second substrate). Following elimination of nitrogen, ring closure could occur via formation of a second Schiff base.

Inspection of the two mechanisms reveals some interesting features. 1) In the first mechanism, Zn^{++} acts via Lewis acid polarization of the carbonyl, while in the second, it acts through nucleophilicity enhancement. Both types of activity have been observed in model and enzymatic reactions using zinc (111). 2) In both, the Schiff base inter-

---

[4]) Proposed by Frank Kung, Department of Molecular Biology, University of California, Berkeley, California.

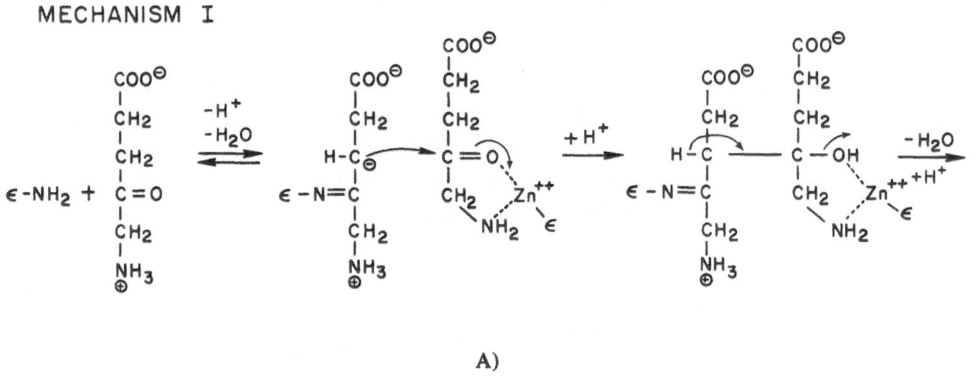

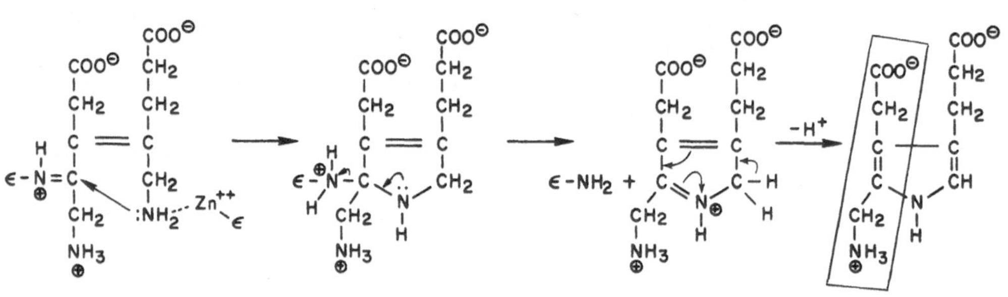

A. M. Cheh and J. B. Neilands

MECHANISM I

A)

Fig. 6. Proposed mechanisms for ALA–D from beef liver.
A) Mechanism I   B) Mechanism II
In both mechanisms the part of PBG originating from the Schiff base bound ALA is boxed.
From (9).

mediate and zinc act *in concert* to catalyze the aldol condensation; this explains how ALA–D *appears* to be both a type I and a type II aldolase. 3) In both, zinc participates in the liberation of H_2O, in concert with the Schiff base in mechanism I, and by catalyzing ring closure through formation of a second Schiff base in mechanism II; hence, it has hydro-lyase activity in addition to aldolase activity. 4) The two mechanisms are mutually exclusive. In mechanism I, carbon-3 of the Schiff base bound substrate must be oriented directly behind the carbonyl of the other substrate to allow nucleophilic attack; in mechanism II, the same type of orientation requirement causes C–3 of the Schiff base bound substrate and the carbonyl of the other substrate to be placed far apart. So, if this type of mechanism is correct, it must be one or the other, but not both. It seems unlikely even if conformational changes took place (suggesting flexibility) that the enzyme would allow both radically different substrate orientations to exist. Aldolases are classified type I or II depending on whether a Schiff base or a metal generates the carbanion. Therefore, ALA–D is either a type I or type II aldolase, but not both. The finding of a Schiff base or a metal requirement has been used

150

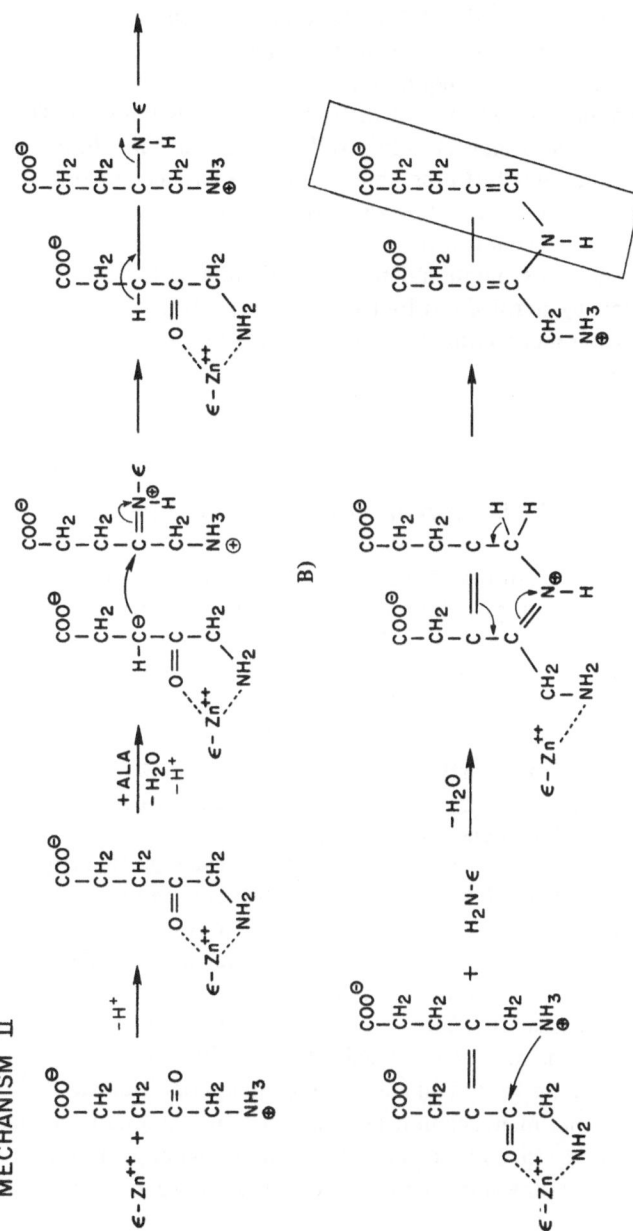

MECHANISM II

to type aldolases as I or II. Where both catalytic functions are found, however, either one or the other may not have anything to do with carbanion formation. Therefore, the finding of either a Schiff base or a required metal may not be sufficiently rigorous proof of an aldolase type until the absence of the other is demonstrated. 5) The two mechanisms are distinguishable from each other. In Fig. 6a and in Fig. 6b, the part of the pyrrole originating from the Schiff base bound substrate has been boxed; as can be seen, this particular substrate molecule becomes one side of the pyrrole in mechanism I and the other side in mechanism II.

A third mechanism, that of separate independent subunits with Schiff base or metal ion catalysis appears to be ruled out by the fact that $NaBH_4$ + ALA reaction eliminates activity in a linear and complete manner, as does metal removal.

C. In vivo Studies — A Zn^{++} Requirement for ALA–D Synthesis

A number of interesting *in vivo* studies of the effect of Zn^{++} on ALA–D activity have been reported. *Komai* and *Neilands* (6) found that growth in Zn^{++} deficient media caused a sharp reduction in *U. sphaerogena* ALA–D. *Finelli et al.* (37) reported a similar adverse effect of a Zn^{++} deficient diet on rat erthrocyte and liver ALA–Ds; they also reported that anemia resulting from Cu^{++} deficiency had no effect on ALA–D levels in rats or in a single human patient.

The addition of Zn^{++} *in vitro* to erthrocyte preparations from Zn^{++} deficient rats (37, 58) or to extracts of Zn^{++} deficient *U. sphaerogena* (6), however, caused no increase in activity, suggesting the absence of apoenzyme.

Abdulla and *Haeger-Aronsen* (61) reported P reliminary evidence that zinc administration to humans causes a rise in ALA–D levels. They (61) and *Hampp* and *Kriebitzsch* (118) reported that assay of ALA–D 4 hours to up to 6 weeks after addition of Zn^{++} to intact erythrocytes also revealed an increase in enzymatic activity.

It was concluded by *Komai* and *Neilands* (6) and *Finelli et al.* (37) that evidence from *in vivo* experiments suggests that Zn^{++} is required for the synthesis of ALA–D. The possibility that the enzyme may be degraded more rapidly during Zn^{++} deficiency needs to be investigated. Failure to find an apoenzyme under conditions of Zn^{++} deficiency means the *in vivo* experiments cannot help identify Zn^{++} as a native metal required by ALA–D for activity (6, 37); however, the finding that copper deficiency does not result in lowered enzyme activity may tend to rule out such a role for copper.

VIII. Two Types of ALA–D?

A. Introduction

Rutter (*116, 117*) has placed fructose-1,6-diphosphate aldolases into one of two classes based on certain characteristics common to the members of each class. Class I aldolases, which are found in animals, plants, protozoa, and green algae, use Schiff base intermediates in catalysis. Class II aldolases, found in bacteria, yeast and other fungi, and blue green algae, have divalent metal ion requirements, monovalent cation activation, and require thiols for maximum activity. Obviously, with the inclusion of fungi (eucaryotes) among the lower organisms, this is not a clear eucaryotic – procaryotic division. He also suggested that other enzymes catalyzing aldol condensations may be segregated into these same two classes. Beef liver ALA–D, which also catalyzes an aldol condensation has both a Schiff base intermediate and a metal-sulfhydryl requirement; therefore, by this definition, it qualifies as both a class I and a class II aldolase (*1, 33*) (see above for a discussion at the mechanistic level for ALA–D). If, as has been suggested, members of a class are likely to exhibit similarities in the active site and catalytic mechanism (*116, 117*), it would be interesting to see if there are common features between the dehydratases and other class I or class II aldolases. Besides the Schiff base intermediate, metal and sulfhydryl requirements, ALA–Ds may also have histidine as an essential catalytic residue, and monovalent cation activation.

B. Procaryotic and Eucaryotic ALA–D

Obviously, ALA–D is a special case among aldolases; because of its unusual nature and because of variation due to evolutionary divergence, it may be difficult to sort out possible homologies with other class I or class II aldolases. Meanwhile, the characteristics of dehydratases derived from different sources suggest a division of ALA–Ds into two classes and possible homology among the members of each class. *Shemin* (*1*) noted that the observation of a different sensitivity to EDTA in comparing the *Rhodopseudomonas* and eucaryotic enzymes, coupled with the initial finding of a Schiff base intermediate in the procaryotes led to speculation that the latter utilized a Schiff base in the enzyme mechanism while eucaryotes relied upon a metal for catalysis. He also noted that such a division would be similar to the aldolase division except procaryotic FDP aldolases use metal ions, while eucaryotic FDP aldolases have Schiff base intermediates – the reverse of the dehydratase situation.

However, all eucaryotic ALA–Ds have been found to utilize Schiff base intermediates, and a mutant *R. spheroides* enzyme exhibits EDTA sensitivity. Therefore, *Nandi* and *Shemin* (*36*) limited speculation about a possible procaryotic-eucaryotic division to the question of whether or not activation by metallic cations was required. Their finding that *Rhodopseudomonas capsulata* ALA–D is similar to the eucaryotic enzyme in not requiring metal ions (36) would also seem to rule out that basis for differentiation.

We now know more about the possible metal requirements of eucaryotic ALA—Ds; in addition, reports on several non-*Rhodopseudomonas* procaryotic ALA—Ds have recently appeared in the literature.(*13, 23, 30, 31*). One can now try again to discuss a basis for distinguishing between procaryotic and eucaryotic ALA—D.

Procaryotic FDP aldolases utilize a divalent transition metal (*116*), while *eucaryotic* ALA—Ds may have the same requirement; the possibility of confusion between the FDP aldolase and ALA—D divisions is apparent. To avoid this, we will call ALA—D either P (procaryotic) type or E (eucaryotic) type rather than type I and type II.

We note in passing that a rough distinction may be made between *P* and *E* ALA—D based on *pH* optima. Generally, the procaryotic and plant enzymes have pH optima above 7.0, often as high as 8.5; the enzyme from eucaryotes generally is most active on the acid side of *pH* 7.0. However, there are exceptions to this generalization (*1, 13, 64*), and differences in pH optimum only indirectly reflect possible molecular and structural differences between the two classes of ALA—D. Therefore, we will confine this discussion to metal ion requirements.

1. Procaryotic

Activity of *Spirillum itersonii* ALA—D is very low without added Mg^{++} or Mn^{++}; K^+ is ineffective (*31*). *Mycobacterium phlei* ALA—D is activated by Mg^{++} and inhibited by EDTA, with reversal of inhibition by Mg^{++}; K^+ is inhibitory (*30*). *Propionibacterium shermanii* ALA—D is also inhibited by EDTA, with reversal by Mg^{++} (*13, 23*).

ALA—D from *R. spheroides* is strongly activated by K^+; however, Mg^{++} is nearly as effective, and Mn^{++} at low concentrations is at least as effective as K^+ (*18*). It is possible that the *R. spheroides* enzyme is a Mg^{++} (Mn^{++}) activated enzyme where monovalent cations may substitute for Mg^{++}. In fact, a mutant *R. spheroides* has been isolated whose dehydratase is inhibited by EDTA, and has a requirement for Mg^{++} (*1*) so, it would seem, particularly if a single point mutation caused the change, that the *R. spheroides* ALA—D is probably very similar to the other Mg^{++} requiring procaryotic dehydratases.

Before going further, one must consider *R. capsulata*; unfortunately, it does not fit in with the other bacterial enzymes. While all the other procaryotic ALA—D forms described thus far have a Mg^{++} (Mn^{++}) or monovalent cation requirement, the *R. capsulata* enzyme does not appear to need any metals (*36*), nor is it inhibited by EDTA. It is assumed that the lack of metal requirement is not due to the *R. capsulata* enzyme having a tightly bound metal which is inaccessible to EDTA; that may require further investigation. Procaryotic ALA—Ds then, seem to either require Mg^{++} (Mn^{++}), or monovalent ions, or may have no metal ion requirements at all.

2. Eucaryotic

Among ALA—Ds from eucaryotic sources, the beef liver enzyme has been the most thoroughly studied regarding the functional metals. Beef liver ALA—D is probably a zinc metalloenzyme. Mg^{++} and Mn^{++} are not found in the purified enzyme, nor are they effective in reconstituting the apoenzyme, so, while one cannot rule out possible roles for copper and iron as functional metals, one probably can do so with Mg^{++} or Mn^{++}. One cannot tell if an enzyme has a transition metal as opposed to a Mg^{++} requirement on the basis of inhibition by EDTA; o-phenanthroline, however, may be of use in making this distinction because of much tighter binding to transition metals than to Mg^{++} (107), providing one avoids the artifactual pitfalls of inhibition due to aromaticity (seen with glutamate dehydrogenase, 92) or sulfhydryl oxidation (seen with rabbit muscle aldolase, 119). By the criterion of o-phenanthroline inhibition, rat liver (33), U. sphaerogena (7), N. crassa (34), and Saccharomyces cerevisiae (26) ALA–Ds probably require transition metals rather than Mg^{++}. Bathocuproinedisulfonate inhibited U. sphaerogena ALA–D is partially reactivated by Zn^{++} (7). More notably, the inactive apodehydratases of N. crassa (34) and S. cerevisiae (26) are reconstituted to an active form by Zn^{++}; therefore, it appears that the lowest eucaryotes, yeasts, molds, and other fungi may have Zn^{++} ALA–D.

At this point, a word of caution is in order. During reconstitution of an apoenzyme, adventitious contamination is hard to control, and an apparent activation of a Zn^{++} requiring apoenzyme may occur upon addition of 10^{-2} M Mg^{++} containing 0.001 % Zn as a contaminant. Furthermore, while some studies on procaryotic ALA–Ds showed transition metals were inhibitory rather than activating, the levels tried were far too high. Generally, 10^{-3} M or greater concentration of Zn^{++} was used; these levels are inhibitory to the beef liver enzyme, (9, 37, 58) and may be so to the enzyme from other sources. So, one cannot as yet completely rule out the possibility that ALA–D in some procaryotic species might require transition metals. However, the findings that monovalent cations (1, 18) and mM concentrations of Mg^{++} (16, 30, 31) are effective, strongly suggest the procaryotic ALA–Ds are metal (Mg^{++}, Mn^{++}, K^+) activated enzymes (expected stability constants $< 10^8 M^{-1}$).

The description of an eucaryotic ALA–D is based primarily on our studies with the beef liver enzyme. Few studies on a possible metal requirement in other eucaryotic ALA–Ds exist; such studies generally have gone no further than to show inhibition by EDTA. Where reversal of EDTA inhibition by added metals was investigated, only a too limited number of metals were tried. In some cases, dialysis to remove EDTA restored a substantial amount of activity, indicating metal ions were picked up from the dialysis buffer. One should not believe that reactivation upon dialysis against Mg^{++} found by Granick and Mauzerall (10) means rabbit reticulocyte ALA–D is a Mg^{++} requiring enzyme; these authors correctly made no such claim.

Without more extensive study, taking particular care to avoid metal contamination or other artifacts, one cannot state that all eucaryotic ALA–Ds are transition metal metalloenzymes. Unfortunately, as noted above, it is no simple matter to determine if an enzyme is metallo or metal activated, or what the native metal is; certainly, it is more difficult than telling whether the mechanism involves metal ions or Schiff base intermediates. Therefore, after pointing out and emphasizing the difficulties involved and the tentative nature of the classification, we state our hypothesis: class P (procaryotic) ALA–Ds are either metal activated enzymes (Mg^{++}, Mn^{++}, monovalent cations), or do not require metal ions; class E (eucaryotic) ALA–Ds are transition metal $(Zn^{++}, Fe^{n+}, Cu^{n+})$ metalloenzymes.

We are proposing a division diagnosed by the nature of the required metal (later transition, probably Zn^{++}, vs. Mg^{++} (Mn^{++}) or monovalent or none) and perhaps the affinity of the enzyme for its metal. The division based on identity of the required metal is strikingly similar to that proposed for superoxide dismutase (SOD) (120). SOD, like ALA–D, also appears to be different from FDP aldolase in that a clear distinction exists between the procaryotic type and eucaryotic type.

Four types of SOD have been found (121, 122), requiring respectively, Mn (intracellular E. coli, and mitochondrial), Fe (periplasmic E. coli) or Cu–Zn (eucaryotes). Since both Fe and Mn SOD enzymes have been found in procaryotes, by analogy there might be several different metal requirements $(Mg^{++}, Mn^{++}, etc.)$ among procaryotic ALA–Ds. Similarly, eucaryotic ALA–Ds might also be of several different metallo types; the validity of the proposed ALA–D division rests on a lack of any metal requirement overlap between procaryotic and eucaryotic enzymes.

Thus far, no mention has been made of plant ALA–Ds. Higher plants (eucaryotes) seem to have EDTA inhibited and Mg^{++} requiring ALA–Ds; this is the case with wheat leaves (16), spinach leaves (123) and tobacco leaves (22). However, this does not invalidate our procaryote-eucaryote division. Plant ALA–D seems to be associated with the chloroplasts (16, 123) or, at least seems to be closely related to chloroplast or chlorophyll synthesis (20, 22, 124, 125). If it is synthesized by the chloroplast, then one would expect to find the procaryotic type of dehydratase. A recent paper suggests the possibility of cytoplasmic synthesis (126); still, since the final product is the magnesium porphyrin chlorophyll, it would not be surprising to find a Mg^{++} dependent or procaryotic type of ALA–D. Furthermore, mitochondrial SOD, which is of the procaryotic type, has been shown to be synthesized outside the mitochondria (121); therefore, synthesis of ALA–D outside of the chloroplast does not negate the likelihood of its being of the procaryotic type.

ALA–D from eucaryotes other than plants appears to be exclusively a cytoplasmic enzyme (127). Virtually all the enzymatic activity is found in the post mitochondrial supernatant. Therefore, studies on ALA–D, unlike those with SOD, will be of no use in helping to determine how mitochondria evolved. However, they may shed light on the evolvement of chloroplasts if a separate cytoplasmic eucaryotic ALA–D can be found in higher plants.

The proposed procaryote – eucaryote division of ALA–D, if proven valid, would place the enzyme in a rather intriguing niche in the scheme of biochemical evolution. Besides having a potential bearing on the origin of chloroplasts, the evolution of ALA–D is also interesting because features of both invariance (Schiff base intermediates) and divergence between procaryotes and eucaryotes (different metal requirements) may be incorporated in the same molecule. Then there remains the question of whether homology exists between ALA–Ds and both types of FDP aldolase, and whether all three enzymes descended from a common aldolase progenitor. It is hoped that this discussion will form a useful basis for further research.

IX. Environmental Aspects of ALA—D

A. Introduction

We have already noted that the mammalian ALA—Ds as thiol requiring enzymes, are quite sensitive to inhibition by heavy metal ions. The enzyme from different tissues exhibits varying sensitivity to metals, the form existing in erythrocytes being very easily poisoned by ingested lead ions.

In general, we shall here be concerned only with the most recent literature pertaining to depression of ALA—D by environmental lead. The subject of atmospheric contamination by lead has been reviewed in a report by the *National Academy of Science* (*128*). A volume, edited by *Griffin* and *Knelson* (*129*), contains extensive contributions from industrial scientists. One chapter in the new book by the late *Henry A. Schroeder* (*130*) is devoted to lead in human health.

B. Natural Lead

Although widely distributed in the lithophere, lead must be classed as a trace element since it averages only about 12 parts per million in the surface rocks. The common lead ore is galena. Mines in Australia, Mexico, Canada and the U.S. supply most of the lead in world industry. Among the salts of Pb^{2+}, the nitrate and acetate are freely soluble but tend to hydrolyze and precipitate the metal hydroxide. Radon gas escaping from the crust of the earth will decay through its various nuclides to isotopes of lead — a kind of continuing natural contamination of the biosphere (*Grandjean, 131*).

C. Industrial Production of Lead

The cupellation process for extracting silver from lead ores, a practice which began about 2500 BC, resulted in the accumulation of large quantities of lead as a byproduct. The stockpiles of waste lead took a quantum jump about 650 BC with the introduction of silver coinage. During the Greek period the waste lead was put to use in roofing, plumbing, anchors, sheathing of various sorts, medical nostrums and in containers for food and beverage (*Patterson* (*132*). In the Roman times lead consumption grew to almost 4 kg/capita/year and this heavy use is believed by some historians to be responsible for the collapse of the empire (*Gilfillan, 133*). Following decay of these earlier civilizations, world lead production rose rapidly with the advent of the industrial revolution and now exceeds 8 billion kg/year. Within the United States the per capita annual consumption is ca. 6 kg. Modern applications include use in storage batteries, solders, paints and as gasoline additives, the latter accounting for some 200 million kg (*Penney et al., 134*).

157

D. Environmental Lead

The introduction of lead tetraethyl a half century ago as the cheapest means of raising the octane rating of gasoline catapulted lead into the status of a major world wide pollutant, with concomitant special hazards to the urban environment. The addition of the better part of a gram of lead tetraethyl per liter can raise the Research Octane Number as much as 10 points and thus prevent "knock" in high compression ratio internal combustion engines. It is estimated that about 2 kg of lead comes out of each automobile exhaust per year, and most of this settles on the roadway and becomes part of the dust, which may contain several thousand parts per million lead (*Schroeder, 130*). Lead salts may be sweet to the taste and young children living in old tenements often acquire the habit of nibbling on paint chips (US observation). The condition of eating extraneous substances is termed "pica", after the Italian for magpie. It is responsible for many of the cases of frank lead poisoning encountered in children in the inner city (*135*). Work performed by the Institute for Local Self Reliance and reported in the New York Times for February 22, 1976, indicated produce grown in urban gardens to be contaminated with significant amounts of lead, although dust was considered to be the major source of the total of 400 μg inhaled and ingested per day. *Goldberg* (*136*) reported that the soft water for drinking that was contained in lead lined tanks in the Scottish Highlands resulted in several cases of poisoning.

One decade ago *Clair C. Patterson* (*137*) published his classic paper on "Contaminated and natural lead environments of man", in which he concluded that residents of the United States are undergoing severe chronic lead insult. This report prompted public health officials to be concerned about the problem of environmental lead.

Recently U.S. Federal courts in response to a suit brought by the Natural Resources Defense Council, have ordered the Environmental Protection Agency to list lead, as is required by the 1970 Clean Air Act, as a pollutant which has particular adverse effects on public health. (New York Times, March 3, 20, 1976).

The human diet may contain several tenths of a milligram of lead per day; the ingestion of 1 mg/day over a certain time period will result in symptoms of toxicity. Although much smaller amounts of lead are inhaled, a greater portion of the substance taken in via this route is incorporated into the body. At about 40 μg of lead per 100 ml of blood, ALA begins to appear in the urine. Small children absorb lead much more efficiently than adults.

A sharp distinction must be made between the long term hazards of poisonous inorganic substances mined and placed in the biosphere by technological man, such as lead, and the toxic synthetic organic chemicals. Even the most durable of the xenobiotics are eroded in decades by solar ultra violet radiation, if not by biodegradation mechanisms. The former, in contrast, are already "mineralized" and will remain as more-or-less permanent pollutants.

E. Logic of the ALA—D Test for Lead Intoxication

While inhibition of blood ALA—D is only one of several tests for lead poisoning, it appears to have several potential advantages over the other methods. It is an exceptionally sensitive indicator of both acute and chronic lead poisoning. Gross clinical symptoms, such as excretion of ALA, do not occur below ca. 40 μg% in the blood. The protoporphyrin assay reflects bone marrow lead and chronic poisoning; in addition, false positive tests appear in the case of iron deficiency anemia (*Granick et al., 41*). Blood is a relatively easy tissue to collect and the ALA—D test does not require excessively expensive reagents and equipment. Admittedly, however, it is not as convenient as the use of porphyrin fluorescence as a rapid screen.

F. Differential Action of Lead on the ALA—D of Various Organs and Tissues

Hammond (138), working with rats, noted that administration of lead lowered the ALA—D activity of both blood and liver, but the former was especially affected. Feeding of EDTA caused slight inhibition of the enzyme. In lead loaded rats the chelating agent caused rapid reactivation of the liver enzyme but was without effect on the enzyme of red cells. *Hampp et al. (118)* showed that erythrocyte ALA—D is somewhat more sensitive to lead ions than is the enzyme of spinach chloroplasts.

G. Reliability of Erythrocyte ALA—D Depression as an Index of Lead Intoxication

Studies in many different laboratories in recent years have confirmed the usefulness of the erythrocyte ALA—D assay as the probe of choice for lead poisoning. The lower levels of enzyme activity seen in patients with the high blood lead levels could, however, have been an artifact of the assay procedure — that is, during the hemolysis step lead could have been redistributed in such a manner that some of the heavy metal ions found their way to the thiol groups of the protein. However, *Maxfield* and *Henry* (139), working with mixed samples of blood from exposed and non-exposed dogs, showed convincingly that ALA—D is actually depressed *in vivo* by lead.

In human subjects not exposed to lead *Secchi et al. (140)* reported considerable scatter in the levels of both liver and erythrocyte ALA—D, which they attributed to differential absorption by individuals.

Systematic analyses of occupationally exposed workers attests to the reliability of the ALA—D test to detect even early exposure to lead (*Hernberg et al., 44; Secchi and Alession, 56*). In general, the enzyme assay was shown to display a better correlation to body lead and to lead absorption than did other physiological parameters such as

159

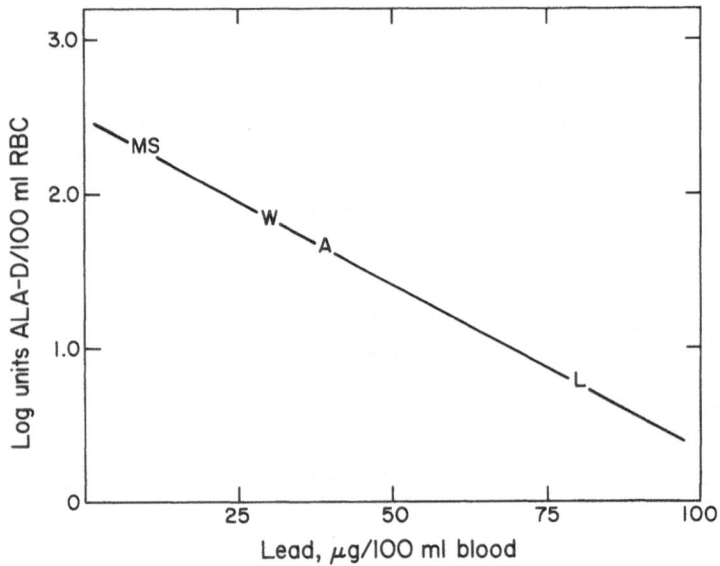

Fig. 7. Negative correlation of blood lead concentration with log activity ALA–D. MS: medical students, W: workers in print shops, A : auto repair workers, L : lead smelters and scrapers. Adapted from Fig. 1 of *Hernberg et al.* (*44*). Basically similar regression curves have been reported in many subsequent investigations. See, for example, *Weissberg et al.* (*141*).

urinary ALA or accumulation of porphyrin in the red cell. (*Sakurai et al.*, *47*; *Franzinelli et al.*, *142*). The latter substance, often referred to as "free erythrocyte protoporphyrin", is in fact the zinc chelate – also fluoroescent (*Lamola* and *Yamane*, *143*).

The level of erythrocyte ALA–D shows a satisfactory negative correlation with the amount of lead found in the urine following chelation therapy with EDTA (*Weitzman et al.*, *144*; *Nieburg et al.*, *145*). *Vitale et al.* (*146*), who also used the EDTA mobilization technique, detected large body burdens of lead in situations where the blood level of the metal was less than 80 μg/100 ml. The authors concluded that blood lead was a highly unreliable measure of occupational exposure.

Patients in different stages of alcoholism were assayed for ALA–D levels in blood (*Magid* and *Hilden*, *147*). The authors proposed that in cirrhosis of the liver, lead is accumulated in that organ at an increased rate, and that alcohol in some way releases the metal ion to the blood.

Several workers have stressed the need to control carefully several technical details of the ALA–D assay. Thus *Tomokuni* (*54*) advocates measurement of the disappearance of ALA rather than appearance of PBG on the grounds that the latter can be converted metabolically to other products in crude systems. *Morgan* and *Burch* (*148*) have noted that, for adequate clinical interpretation, the pH should be well controlled. *Kniep et al.* (*60*) showed

160

that large potential errors are involved in the choice of sample collecting tubes. They recommend a container which will neither inactivate nor reactivate the enzyme; only polystyrene was found to be entirely satisfactory. Meaningful interpretation of epidemiological studies requires standardization of the time interval between collection of the sample and its assay; proper storage temperature was also evaluated and established (*Prpic-Majic et al., 46*).

Rapid and convenient though the erythrocyte porphyrin assay may be, it should be combined with ALA—D assay for distinction between lead poisoning and iron deficiency anemia (*Sassa et al., 149*).

H. Specificity of Lead as an Inhibitor of ALA—D

Relatively little attention has been paid to ALA—D activity in hematological disorders. *Lyberatos et al. (150)* showed that erythrocyte ALA—D activity is increased in homozygous B-thalassemia. In peripheral blood of rats, the enzyme activity was found to be directly proportional to the reticulocyte count (*151*).

The stimulatory action of vitamin E on heme synthesis has been ascribed to its effect on δ-aminolevulinate synthetase rather than on ALA—D (*Bartlett et al., 152*). Interestingly, a random sample of hospitalized persons was found to have a significantly lower than normal level of erythrocyte ALA—D (*Bodlaender et al., 153*). *Sassa et al. (154)* concluded that the 4-fold range of erythrocyte ALA—D potency in the normal population probably represents genetic variation, with the activity inherited in a codominant mode.

As a thiol enzyme, isolated ALA—D would not be expected to be uniquely sensitive to inhibition by a single metal ion. Thus many *in vitro* experiments have shown that in addition to lead, mercury, silver, cobalt, cadmium, copper, magnesium — even excess zinc — depress activity. It does seem that *in vivo* the enzyme is most sensitive to lead. *Meredith et al. (155)*, working with rats, reported an *in vitro* activation of erythrocyte ALA—D by low concentrations of aluminum, whereas higher concentrations of the metal ion were inhibitory. Inhibition by lead and aluminum together was demonstrated to be additive. *Lauwerys et al. (52)* investigated a number of clinical parameters in 77 workers occupationally exposed to cadmium; compared to a like number of controls the former group had no correlation between blood cadmium and ALA—D.

I. Properties of Native and Lead Inhibited ALA—D

Tomokuni (55; 156; 157) confirmed earlier reports that the pH optimum of ALA—D from the blood of persons occupationally exposed to lead is shifted from the normal value of 6.4—6.8 to about 5.8—6.0. Heating hemolysates of the blood of lead workers to 60 °C for five minutes increased the ALA−D activity 3.6 fold and shifted the pH optimum back to 6.6. Incubation in the presence of dithiothreitol also restored the higher pH optimum. In normal blood the heat treatment increased activity only slightly and did not alter the value of the pH optimum. In a survey of standard and intoxicated populations, *Dingeon et al.* (*158*) independently described a very significant difference in ALA−D activity at pH 6.9 but not at pH 5.9. These data suggest that a variety of manipulations are effective in removing the lead from the active center of the enzyme.

J. Mechanism of Inhibition of ALA−D by Lead

Granick et al. (*41*) concluded that 20 mM dithiothreitol would fully reactivate the ALA−D in lead poisoned blood, suggesting that the heavy metal ion does not block synthesis or irreversibly destroy the enzyme. Since lead raised the K_m only slightly but lowered substantially the V_{max}, inhibition was concluded to be primarily of the uncompetitive type. In rats exposed to lead, *Finelli et al.* (*58*) found that zinc could reactivate the ALA−D to the levels of controls and the authors suggested a competition between lead and zinc ions for binding to the enzyme. It seems to be agreed that, upon termination of exposure, recovery of ALA−D activity will precede a detectable drop in blood lead values (*Haeger-Aronsen et al.*, 53; *Maxfield et al.*, 48).

K. Effects of Low Levels of Lead on Neurological Functions

While our chief focus has been on the role of lead as an inhibitor of various enzymes in the heme biosynthesis pathway, with special emphasis on its effect upon ALA−D, the saturnine metal is known to interfere with a number of bodily functions. The general physiology of acute and chronic lead intoxication has been reviewed elsewhere (*128*). Recently, attention has been directed to the effect of low levels of lead on the function of the central nervous system. Administration of the substance to young rats increased motor activity and brain norepinephrine (*159*). An epidemiological survey carried out in Glasgow implicated lead in drinking water as the cause of mental retardation in children (*160*). Children and the offspring of pregnant women living in areas with 800 or more μg of lead per liter were especially at risk; the World Health Organization considers 100 μg per liter to be the upper limit for lead contamination of potable water. These studies underline the importance of a reliable, early index of exposure to lead, such as ALA−D activity of erythrocytes.

L. Significance of Lead Depression of ALA–D Activity in Ecology and Public Health

Several investigators have examined the blood of wild animals living in rural and urban environments in order to ascertain the lead contamination of the immediate surroundings. *Ohi et al.* (*161*) advanced the proposal that pigeons could be used as a handy sensor of urban lead pollution and *Mouw et al.* (*59*) found lowered levels of ALA–D in various tissues, including blood, of urban rats.

Heme is required for three important biochemical chores, namely, oxygen transport, electron transport and oxygenation reactions. Chlorophyll compounds are involved in photosynthetic reactions in all green plant tissues, vitamin B_{12} is required for rearrangement of carbon skeletons and, in some species, for reduction of ribotides to deoxyribotides. The biosynthesis of all three tetra-pyrrole-derived substances, namely, heme, chlorophyll and vitamin B_{12}, proceeds through the lead sensitive enzyme, ALA–D; in addition, other steps in heme biosynthesis, such as ferrochelatase, are inhibited by lead. This raises the question of the ultimate hazard of environmental lead. The metal is obviously interfering with some of the most crucial and fundamental reactions which occur in the biosphere. Paradoxically, the metabolic status of ALA–D in erythrocytes is in doubt. Presumably the enzyme plays a limited role in adult red cells. However, should the lead poisoning invoke a deficit of heme for synthesis of enzymes such as cytochrome P–450, the consequences could be profoundly important in biology. Oxygenases may detoxify or potentiate the toxicity of drugs. In a study of lead poisoned rats, *Scoppa et al.* (*162*) found impaired formation of mixed function oxidase, which they attributed to defective heme synthesis. In general, hydroxylation promotes detoxification since it renders drugs more water soluble and facilitates their excretion.

Lead is an element with no known biological function – it is only toxic. There is doubt that there is any threshold below which no biological damage occurs, and the arbitrary figures of 40 or 60 μg/100 ml as upper limits for blood lead should probably be discarded or revised downwards (*Waldron, 163*). Although the lead industry favors the "dead bodies" criterion of toxicity, a more conservative approach would be to accept depressed levels of red cell ALA–D as the signal to search for, and eliminate, the origin of the lead insult (*130*). With these considerations in mind, the California Air Resources Board on January 15, 1976, adopted Resolution 76–2 setting 1.5 μg/m^3 as the limit for airborne lead. Almost a decade earlier, *Schroeder* and *Tipton* (*164*) had concluded "In view of the steadily increasing annual pollution of air and soils with lead from motor vehicle exhausts, inate toxicity in exposed human beings may appear."

X. Summary

ALA–D, the second enzyme in the heme biosynthetic pathway forms porphobilinogen by condensing two molecules of ALA with the loss of two molecules of water.

Enzyme assays based on those of either Gibson or Granick are satisfactory, but potassium ion is incompatible with the Granick method. Assay of erythrocyte ALA–D in sodium phosphate rather than in bicarbonate is advised.

Various purification procedures described in the literature are cited but not discussed.

Beef liver ALA–D has a molecular weight of 280,000–290,000 and a subunit (probably identical) weight of about 35,000, it is an octamer rather than the previously reported hexamer. Only 4 Schiff base sites may be labelled with ALA + $NaBH_4$. Therefore, octameric ALA–D may exhibit half site reactivity; however, one cannot rule out ALA–D having only 4 active sites instead of the presumed 8. A model for the quaternary structure of ALA–D is presented.

ALA–D contains essential thiol(s), and may have a catalytic histidine residue. A limited activation by NH_4^+ may be observed; activation by divalent metal ions is in doubt. Excesses of Zn^{++}, Co^{++}, Cu^+, Cu^{++}, Hg^{++}, and Pb^{++} are lethal to ALA–D.

Metalloenzymes are defined through determining the identity, stoichiometry and strength of binding of bound metals, and demonstrating their essentiality. Inhibition of ALA–D by a diverse set of chelators, with reversal of inhibition by metal ions is seen; Zn^{++} appears to be the most effective in restoring activity, but other metal ions are nearly as functional. If purified ALA–D is regarded as in its native state, it could be misinterpreted to be a one Zn^{++} enzyme. An apoenzyme may be created with EDTA; it does not bind the chelator tightly. Measurement of activity using high concentrations of apoenzyme at $0\ ^{\circ}C$, to avoid regeneration of activity by adventitious contaminants, indicates that at least 5–6 Zn^{++} are required for activity. Other physiological metals are $< 25\%$ as effective as Zn^{++}. It is felt that ALA–D is probably an eight Zn^{++} enzyme; roles for Cu or Fe, however, cannot be ruled out. Spectra obtained after Cd^{++} replacement of Zn^{++} in ALA–D indicate thiols may be among the metal binding ligands. The difficulties and possible misconceptions arising from attempted application of the metalloenzyme defining criteria to ALA–D are discussed.

No evidence exists that either favors or opposes a structural or substrate binding role for Zn^{++} in ALA–D. Still, by analogy with other Zn^{++} enzymes, a catalytic role is proposed. Two detailed and distinguishable enzyme mechanisms are put forth; in both, Zn^{++} has aldolase and lyase activity. Besides having a possible catalytic role in ALA–D, Zn^{++} appears to be required *in vivo* for synthesis of the enzyme.

The FDP aldolase division of Rutter is discussed. A similar type of division of ALA–D is proposed; it is suggested that type P (procaryotic) ALA–Ds require Mg^{++} (Mn^{++}) or monovalent cations, or have no metal requirements, while type E (eucaryotic) ALA–Ds require a group VIII or IB or IIB transition metal. Procaryotic ALA–Ds may also bind metals less tightly than eucaryotic ALA–Ds. The ALA–D division is a clear procaryote-eucaryote one; thus it differs from the FDP aldolase division, but is similar to that suggested for SOD. The potential interest in charting the biochemical evolution of ALA–D is discussed.

Of the several enzymes in the heme biosynthesis pathway inhibited by lead, ALA–D is the most sensitive to environmental exposure to this metal. Erythrocyte ALA–D is depressed *in vivo* by lead and the lower enzyme activity measured in lead poisoned individuals is not an artifact introduced by the hemolysis procedure with concomitant redistribution of lead in the sample.

All of the available evidence suggests that the ALA–D level of erythrocytes is a specific and sensitive indicator of body burden of lead. As an index of intoxication with this heavy metal it is superior to measurement of urinary ALA or erythrocyte porphyrin. ALA–D inhibited by lead has a lower pH optimum than the native enzyme. The activity and the pH optimum of the lead depressed enzyme can be restored to normal by various manipulations, including heating and treatment with thiol compounds.

It is concluded that mild depression of ALA–D activity, while not in itself a proven health hazard, is the subclinical signal to begin reduction of airborne and other forms of environmental exposure to lead.

Notes added in proof

Shemin (*1*) has reiterated that ALA–D has a molecular weight of 285,000 daltons, a subunit weight of 35,000, and is composed of eight subunits, only four of which form Schiff bases with substrate at any one time. He also reports the following: lysine residues are responsible for Schiff base formation. Reduced thiols are required for enzymatic activity, and of the two rapidly reacting cysteines out of a total of seven in each subunit, one may be protected by substrate. Four to six Zn^{++} are found in each mole of crystallized enzyme, confirming our titration data described above. ALA–D from liver and *R. spheroides* is quite stable when bound to a solid support. *R. spheroides* ALA–D may be dissociated into tetramers with retention of activity; it remains to be seen if the bacterial enzyme exhibits the same features of subunit interaction such as half site reactivity that are observed with beef liver ALA–D.

Levitzki and *Koshland* (*2*) have published a comprehensive review of negative cooperativity and half-of-the-sites reactivity which sets forth their viewpoint of the subject, as opposed to that of *Seydoux, et al.* (*3*).

1. *Shemin, D.:* Phil. Trans. R. Soc. London *B273*, 109 (1976).
2. *Levitzki, A., Koshland, D. E., Jr.:* In: Current Topics in Cellular Regulation (Horecker, B. L. and Stadtman, E. R., eds.) Vol. 10, p. 1. New York: Academic Press 1976.
3. *Seydoux, F., Malhotra, O. P., Bernhard, S. A.:* CRC Crit. Revs. Biochem. *2*, 227 (1974).

Acknowledgements. This review is derived, in part, from the doctoral dissertation of *Albert M. Cheh* (*9*) and was supported by USPHS grant No. AI–04156.

References

1. *Shemin, D.:* In: The Enzymes (Boyer, P. D., ed.) 3rd ed., Vol. 7, p. 323. New York: Academic Press, 1972.
2. *Nandi, D. L., Shemin, D.:* J. Biol. Chem. *243*, 1236 (1968).
3. *Gibson, K. D., Neuberger, A., Scott, J. J.:* Biochem. J. *61*, 618 (1955).
4. *Batlle, A. M., Ferramola, A. M., Grinstein, M.:* Biochem. J. *104*, 244 (1967).
5. *Tomio, J. M., Tuzman, V., Grinstein, M.:* Eur. J. Biochem. *6*, 84 (1968).
6. *Komai, H., Neilands, J. B.:* Arch. Biochem. Biophys. *124*, 456 (1968).
7. *Komai, H., Neilands, J. B.:* Biochem. Biophys. Acta *171*, 311 (1969).
8. *Cheh, A., Neilands, J. B.:* Biochem. Biophys. Res. Comm. *55*, 1060 (1973).
9. *Cheh, A.:* PhD. Dissertation, University of California, Berkeley, California, 1974.
10. *Granick, S., Mauzerall, D.:* J. Biol. Chem. *232*, 119 (1958).
11. *Mauzerall, D., Granick, S.:* J. Biol. Chem. *219*, 435 (1956).
12. *Russell, R. L., Coleman, D. L.:* Genetics *48*, 1033 (1963).
13. *Walerych, W.:* Acta Biochim. Polon. *10*, 243 (1963).
14. *Calissano, P., Cartasegna, C., Matteini, M.:* Ital. J. Bioch. *15*, 18 (1966).
15. *Coleman, D. L.:* J. Biol. Chem. *241*, 5511 (1966).
16. *Nandi, D. L., Waygood, E. R.:* Can J. Biochem. *45*, 327 (1967).
17. *Nandi, D. L., Baker-Cohen, K. F., Shemin, D.:* J. Biol. Chem. *243*, 1224 (1968).
18. *Nandi, D. L., Shemin, D.:* J. Biol. Chem. *243*, 1231 (1968).
19. *Doyle, D., Schimke, R. T.:* J. Biol. Chem. *244*, 5449 (1969).
20. *Steer, B. T., Gibbs, M.:* Plant Physiol. *44*, 781 (1969).
21. *Shetty, A. S., Miller, G. W.:* Biochim. Biophys. Acta *185*, 458 (1969).
22. *Shetty, A. S., Miller, G. W.:* Biochem. J. *114*, 331 (1969).
23. *Muller, G., Bezold, G.:* Z. Naturforsch. *24B*, 47 (1969).
24. *Muthukrishnan, S., Padmanaban, G., Sarma, P. S.:* J. Biol. Chem. *244*, 4241 (1969).
25. *Tigier, H. A., Batlle, A. M. del C., Locascio, G. A.:* Enzymologia, *38*, 43 (1970).
26. *Labbe, P.:* Biochimie *53*, 1001 (1971).
27. *Nandi, D. L.:* Arch. Biochem. Biophys. *142*, 157 (1971).
28. *Doyle, D.:* J. Biol. Chem. *246*, 4965 (1971).
29. *Van Heyningen, S., Shemin, D.:* Biochem. *10*, 4676 (1971).
30. *Yamasaki, H., Moriyana, T.:* Biochim. Biophys. Acta *227*, 698 (1971).
31. *Ho, Y. K., Lascelles, J.:* Arch. Biochem. Biophys. *144*, 734 (1971).
32. *Wilson, E. L., Burger, P. E., Dowdle, E. B.:* Eur. J. Biochem. *29*, 563 (1972).
33. *Gurba, P. E., Sennett, R. E., Kobes, R. D.:* Arch. Biochem. Biophys. *150*, 130 (1972).
34. *Muthukrishnan, S., Malathi, K., Padmanaban, G.:* Biochem. J. *129*, 31 (1972).
35. *Yamada, M.:* Biochim. Biophys. Acta *279*, 535 (1972).
36. *Nandi, D. L., Shemin, D.:* Arch. Biochem. Biophys. *158*, 305 (1973).
37. *Finelli, V. N., Murthy, L., Peirano, W. B., Petering, H. G.:* Biochem. Biophys. Res. Comm. *60*, 1418 (1974).
38. *Burnham, B. F., Lascelles, J.:* Biochem. J. *87*, 462 (1963).
39. *Granick, S., Sassa, S., Granick, J. L., Levere, R. D., Kappas, A.:* Proc. Nat. Acad. Sci., *69*, 2381 (1972).
40. *Weissberg, J. B., Voytek, P.:* Biochim. Biophys. Acta, *364*, 304 (1974).
41. *Granick, J. L., Sassa, S., Granick, S., Levere, R. D., Kappas, A.:* Biochem. Med. *8*, 149 (1973).
42. *Bonsignore, D., Calissano, P., Cartasegna, C.:* Med. Lavoro, *56*, 199 (1965).
43. *DeBruin, A., Holboom, H.:* Brit. J. Indust. Med., *24*, 203 (1967).
44. *Hernberg, S., Nikkanen, J., Mellin, G., Lilius, H.:* Arch. Environ. Health *21*, 140 (1970).
45. *Haeger-Aronsen, B., Abdulla, M., Fristedt, B.:* Arch. Environ. Health, *23*, 440 (1971).
46. *Prpic-Majic, D., Mueller, P. K., Lew, V. C., Twiss, S.:* Amer. Indust. Hygiene J. *34*, 315 (1973).
47. *Sakurai, H., Sugita, M., Tsuchiya, K.:* Arch. Environ. Health, *29*, 157 (1974).

48. *Maxfield, M. E., Stopps, G. J., Barnes, J. R., Snee, R. D., Finian, M., Azar, A.:* Amer. Industr. Hyg. Assoc. J. *36*, 193 (1975).
49. *Nakao, K., Wada, O., Yano, Y.:* Clin. Chim. Acta., *19*, 319 (1968).
50. *Collier, H. B.:* Clin. Biochem. *4*, 222, (1971).
51. *Hernberg, S., Tola, S., Nikkanen, J., Valkonen, S.:* Arch. Environ. Health, *25*, 109 (1972).
52. *Lauwerys, R. R., Buchet, J. P., Roels, H. A.:* Brit. J. Indust. Med. *30*, 359 (1973).
53. *Haeger-Aronsen, B., Abdulla, M., Fristedt, B.:* Arch. Environ. Health, *29*, 150 (1974).
54. *Tomokuni, K.:* Clin. Chem. *20*, 1287 (1974).
55. *Tomokuni, K.:* Arch. Environ. Health, *29*, 274 (1974).
56. *Secchi, G. C., Alessio, L.:* Arch. Environ. Health, *29*, 351 (1974).
57. *Taniguchi, N., Sato, T., Kondo, T., Tanachi, T., Saito, K., Takakuwa, E.:* Clin. Chim. Acta *59*, 29 (1975).
58. *Finelli, V. N., Klauser, D. S., Karaffa, M. A., Petering, H. G.:* Biochem. Biophys. Res. Comm. *65*, 303 (1975).
59. *Mouw, D., Kalitis, K., Anver, M., Schwartz, J., Constan, A., Hartung, R., Cohen, B., Ringler, D.:* Arch. Env. Health *30*, 276 (1975).
60. *Kneip, T. J., Cohen, N., Rulon, V.:* Analyt. Chem., *46*, 1863 (1974).
61. *Abdulla, M., Haeger-Aronsen, B.:* Enzyme *12*, 708 (1971).
62. *Thiers, R. E.:* Meth. Biochem. Anal. *V.*, 273 (1957).
63. *Wu, W. H., Shemin, D., Richards, K. E., Williams, R. C.:* Proc. Nat. Acad. Sci. *71*, 1767 (1974).
64. *DeBarreiro, O. L. C.:* Biochim. Biophys. Acta *139*, 479 (1967).
65. *Wu, W. H., Shemin, D., Richards, K. E., Williams, R. C.:* Fed. Proc. *33*, 1478 (1974).
66. *Cheh, A., Neilands, J. B.:* Fed. Proc., *33*, 1245 (1974).
67. *Tsukamoto, I., Yoshinaga, T., Sano, S.:* Biochem. Biophys. Res. Common. *67*, 294 (1975).
68. *Shemin, D.:* Annals N. Y. Acad. Sci., *244*, 348 (1975).
69. *MacQuarrie, R. A., Bernhard, S. A.:* J. Mol. Biol., *55*, 181 (1971).
70. *Levitzki, A., Stallcup, W. B., Koshland, D. E., Jr.:* Biochem. *10*, 3371 (1971).
71. *Seydoux, F., Malhotra, O. P., Bernhard, S. A.:* CRC Crit. Revs. Biochem. *2*, 227 (1974).
72. *Lazdunski, M.:* In: Current Topics in Cellular Regulation, Vol. *6*, p. 267 (Horecker, B. L., Stadtman, E. R., eds.). New York: Academic Press 1972.
73. *Lazdunski, M.:* Prog. Bioorg. Chem. *3*, 81 (1974).
74. *Woodfin, B. M.:* Biochem. Biophys. Res. Commun. *29*, 288 (1967).
75. *Lai, C. Y., Horecker, B. L.:* Essays Biochem. *8*, 149 (1972).
76. *Komai, H.:* Ph. D. Dissertation, University of California, Berkeley, California, 1968.
77. *Komai, H., Neilands, J. B.:* Science *153*, 751 (1966).
78. *Yasukochi, Y., Nakamura, M., Minakami, S.:* Biochem. J., *144*, 455 (1974).
79. *Maines, M. D., Kappas, A.:* J. Biol. Chem. *250*, 4171 (1975).
80. *Vallee, B. L.:* Advan. Protein Chem. *10*, 317 (1955).
81. *Malmström, B. G., Neilands, J. B.:* Ann. Rev. Biochem. *33*, 331 (1964).
82. *Mildvan, A. S.:* In: The Enzymes (Boyer, P. D., ed.), 3rd ed., Vol. *II*, p. 445, New York: Academic Press 1970.
83. *Scrutton, M. C.:* In: Inorganic Biochemistry, Vol. *1*, chapter 14. (Eichhorn, G. L., ed.). New York: Elsevier 1973.
84. *Lindskog, S., Nyman, P. O.:* Biochim. Biophys. Acta *85*, 462 (1964).
85. *Piras, R., Vallee, B. L.:* Biochem. *6*, 348 (1967).
86. *Malmström, B. G.:* In: The Enzymes, 2nd ed., Vol. *5*, p. 471 (Boyer, P. D., Lardy, H., Myrback, K., eds.). New York: Academic Press 1961.
87. *Vallee, B. L., Rupley, J. A., Coombs, T. L., Neurath, H.:* J. Biol. Chem. *235*, 64 (1960).
88. *Fee, J. A.:* J. Biol. Chem. *248*, 4229 (1973).
89. *Simpson, R. T., Vallee, B. L.:* Biochem. 7, 4343 (1968).
90. *Kobes, R. D., Simpson, R. T., Vallee, B. L., Rutter, W. J.:* Biochem. *8*, 585 (1969).
91. *Adelstein, S. J., Vallee, B. L.:* J. Biol. Chem. *234*, 824 (1959).
92. *Yielding, K. L., Tomkins, G. M.:* Biochim. Biophys. Acta *62*, 327 (1962).

167

93. *Iodice, A. A., Richert, D. A., Schulman, M. P.:* Fed. Proc. *17*, 248 (1958).
94. *Wilson, M. L., Iodice, A. I., Shulman, M. P., Richert, D. A.:* Fed. Proc. *18*, 352 (1959).
95. *Drum, D. E., Li, T. K., Vallee, B. L.:* Biochem. *8*, 3873 (1969).
96. *Applebury, M. L., Coleman, J. E.:* J. Biol. Chem. *244*, 308 (1969).
97. *Csopak, H., Szajn, H.:* Arch. Biochem. Biophys. *157*, 374 (1973).
98. *Csopak, H., Szajn, H.:* Biochim. Biophys. Acta *258*, 466 (1972).
99. *Malmström, B.:* Proc. 8th Meeting FEBS *29*, 119 (1972).
100. *Lindskog, S.:* Structure and Bonding, Vol. *8*, p. 153. Berlin–Heidelberg–New York: Springer 1970.
101. *Vallee, B. L.:* In: Advances in Experimental Biology and Medicine, Vol. *40*, p. 1. (Dhar, S. K., ed.). New York: Plenum Press 1973.
102. *Kägi, J. H. R., Vallee, B. L.:* J. Biol. Chem. *236*, 2435 (1961).
103. *Drum, D. E., Vallee, B. L.:* Biochem. Biophys. Res. Comm. *41*, 33 (1970).
104. *Branden, C. I., Jornvall, H., Eklund, H., Furugren, B.:* In: The Enzymes, 3rd Edition, Boyer, P. D., Editor, XI, *part A*, p. 104, Academic Press, N. Y.
105. *Rosenbusch, J. P., Weber, K.:* Proc. Nat. Acad. Sci. U. S. *68*, 1019 (1971).
106. *Lin, Y. N., Nakamura, S., Kobes, R. D., Kimura, T.:* Biochem. Biophys. Res. Comm. *47*, 1209 (1972).
107. *Vallee, B. L., Wacker, W. E. C.:* In: The Proteins, (Neurath, H., ed.), 2nd ed., Vol. *V*, Academic Press, New York, 1970.
108. *Curdel, A., Iwatsubo, I.:* FEBS Lett, *1*, 133 (1968).
109. *Scrutton, M. C., Griminger, P., Wallace, J. C.:* J. Biol. Chem. *247*, 3305 (1972).
110. *Malmstrom, B. G., Rosenberg, A.:* Adv. Enzymol. *21*, 131 (1959).
111. *Dunn, M. F.:* Structure and Bonding, Vol. *23*, p. 61. Berlin, Heidelberg, New York: Springer 1975.
112. *Drum, D. E., Harrison, J. H., IV, Li, T-k, Bethune, J. L., Vallee, B. L.:* Proc. Nat. Acad. Sci. *57*, 1434 (1967).
113. *Nelbach, M. E., Pigiet, V. P., Jr., Gerhart, J. C., Schachman, H. K.:* Biochem. *11*, 315 (1972).
114. *Rosenbusch, J. P., Weber, K.:* J. Biol. Chem. *246*, 1644 (1971).
115. *Colman, R. F., Foster, D. S.:* J. Biol. Chem. *245*, 6190 (1970).
116. *Rutter, W. J.:* Fed. Proc. *23*, 1248 (1964).
117. *Rutter, W. J.:* In: Evolving Genes and Proteins, (Bryson, V., Vogel, H. J., eds.), p. 279. New York: Academic Press 1965.
118. *Hampp, R., Kriebitzch, C.:* Z. Naturforsch. *30*, 434 (1975).
119. *Kobashi, K., Horecker, B. L.:* Arch. Biochem. Biophys. *121*, 178 (1967).
120. *Fridovich, I.:* In: Molecular Mechanisms of Oxygen Activation, p. 453. New York: Academic Press 1974.
121. *Fridovich, I.:* Adv. Enzymol. *41*, 35 (1974).
122. *Fridovich, I.:* Amer. Sci. *61*, 54 (1975).
123. *Schneider, H. A. W.:* Z. Pflanzenphysiol. *62*, 328 (1970).
124. *Beale, S. I.:* Plant Physiol. *48*, 316 (1971).
125. *Stobart, A. K., Thomas, D. R.:* Phytochem. 7, 1313 (1968).
126. *Schneider, H. A. W.:* Biochem. Biophys. Res. Comm. *60*, 468 (1974).
127. *Lascelles, J.:* Tetrapyrrole Biosynthesis and its Regulation. New York: W. A. Benjamin, Inc. 1964.
128. *Airborne Lead in Perspective:* National Research Council, Washington, D. C.: National Academy of Sciences 1972.
129. *Griffin, T. B., Knelson, J. H. (ed):* Lead. Stuttgart: Georg Thieme 1975.
130. *Schroeder, H. A.:* The poisons around us. Bloomington: Indiana University Press 1974.
131. *Grandjean, P.:* In: Lead, p. 6 (eds. Griffin, Knelson). Stuttgart: Georg Thieme 1975.
132. *Patterson, C. C.:* Connecticut Medicine *35*, 347 (1971).
133. *Gilfillan, S. C.:* J. Occup. Med. 7, 53 (1965).

134. *Penney, D. G., Bederka, J. P., McLellan, J. S., Coello, W. F., Saleem, Z. A., Khan, M. A. Q.:* In: Survival in Toxic Environments, p. 497 (eds. Khan, Bederka). New York: Academic Press 1974.
135. *Chisolm, J. J., Jr.:* Med. Times 98, 92 (1970).
136. *Goldberg, A.:* Environ. Health Perspectives 7, 103 (1974).
137. *Patterson, C. C.:* Arch. Environ. Health 11, 344 (1965).
138. *Hammond, P. B.:* Toxicol. and Appl. Pharm. 26, 466 (1973).
139. *Maxfield, M. E., Henry, N. W.:* In: Lead, p. M (eds. Griffin, Knelson). Stuttgart: Georg Thieme 1975.
140. *Secchi, G. C., Erba, L., Cambiaghi, G.:* Arch. Environ Health 28, 130 (1974).
141. *Weissberg, J. B., Lipschutz, F., Oski, F. A.:* New Eng. J. Med. 284, 565 (1971).
142. *Franzinelli, A., Battista, G., Benelli, A., Sartorelli, E.:* Medicina del Lavoro 65, 206 (1974).
143. *Lamola, A. A., Yamane, T.:* Science 186, 936 (1974).
144. *Weitzman, M. L., Post, E. M., Schneider, T. R., Oski, B. F., Oski, F. A.:* Pediatric Research 9, 285 (1975).
145. *Nieburg, P. I., Weiner, L. S., Oski, B. F., Oski, F. A.:* Am. J. Dis. Child. 127, 348 (1974).
146. *Vitale, L. F., Joselow, M. M., Wedeen, R. P., Pawlow, M.:* J. Occup. Med. 17, 155 (1975).
147. *Magid, E., Hilden, M.:* Int. Arch. Occup. Hlth. 35, 61 (1975).
148. *Morgan, J. M., Burch, H. B.:* J. Lab. Clin. Med. 85, 746 (1975).
149. *Sassa, S., Granick, S., Kappas, A.:* Ann. N. Y. Acad. Sci. 244, 419 (1975).
150. *Lyberators, C., Mitsiou, Ch., Phillippidou, Aik., Papayannis, A. G., Chalevelakis, G., Gardikas, C.:* Scand. J. Haemat. 12, 81 (1974).
151. *Davis, J. R., Messmore, H. L., Fareed, J.:* Fed. Proc. 34, 805 (1975).
152. *Bartlett, R. S., Rousseau, J. E., Jr., Frier, H. I., Hall, R. C., Jr.:* J. Nutr. 104, 1637 (1974).
153. *Bodlaender, P., Ulmer, D. D., Vallee, B. L.:* Analyt. Biochem. 58, 500 (1974).
154. *Sassa, S., Granick, S., Bickers, D. R., Levere, R. D., Kappas, A.:* Enzyme 16, 326 (1973).
155. *Meredith, P. A., Moore, M. R., Goldberg, A.:* Biochem. Soc. Trans. 2, 1243 (1974).
156. *Tomokuni, K.:* Arch. Environ. Health 30, 148 (1975).
157. *Tomokuni, K.:* Arch. Environ. Health 30, 317 (1975).
158. *Dingeon, B., Roullet, A., Prost, G., Tolot, F.:* Arch. Environ. Health 30, 317 (1975).
159. *Golter, M., Michaelson, I. A.:* Science 187, 359 (1975).
160. *Beattie, A. D., Moore, M. R., Goldberg, A., Finlayson, M., Graham, J., Mackie, E., Main, J., McLaren, D. A., Murdoch, R. M., Stewart, G. T.:* The Lancet 1, 589 (1975).
161. *Ohi, G., Seki, H., Akiyama, K., Yagyu, H.:* Bull, Environ, Contam. & Toxicol. 12, 92 (1974).
162. *Scoppa, P., Roumengous, M., Penning, W.:* Experientia 29, 970 (1973).
163. *Waldron, H. A.:* Arch. Environ. Health 29, 271 (1974).
164. *Schroeder, H. A., Tipton, I. H.,* Arch. Environ. Health 17, 965 (1968).

The Biological Chemistry of Gold:
A Metallo-Drug and Heavy-Atom Label
with Variable Valency

Peter J. Sadler

Department of Chemistry, Birkbeck College, University of London, Malet Street, London WC1E 7HX.

Table of Contents

1. Introduction

Current areas of interest in the biological chemistry of gold include:

(i) *Chemotherapy:* (a) as anti-inflammatory agents — Au[I]thiomalate and thioglucose are used clinically for treatment of rheumatoid arthritis.
   (b) [Au[III](5-diazouracil)$_2$Cl$_2$]Cl has anti-tumour activity in mice.
(ii) *Labelling:* gold is a conveniently-heavy atom for labelling macromolecules and whole cells for viewing by electron microscopy and X-ray crystallography.
(iii) *Bacteriocides:* many gold compounds are powerful disinfectants.
(iv) *Obesity:* Au[I]thioglucose induces obesity in mice.
(v) *Ore dissolution:* certain microorganisms act on Au deposits.
(vi) *Plant uptake:* in regions rich in Au deposits.
(vii) *Tissue destruction:* by radioactive, colloidal [198]Au.

We shall focus attention on (i) and (ii), and seek a molecular basis for the action of gold in these areas. It is only when this has been understood that the design of drugs, cited in (i), can be improved, to give increased pharmacological efficiency and reduced toxicity, and the specific labelling in (ii) achieved. This will involve us in a discussion of the molecular structure of gold complexes and their reactivity toward ligand displacement reactions and redox processes.

At times we shall find our knowledge of the basic chemistry of gold inadequate to answer some of the questions which we shall pose as a result of examining biological problems. For example, does H$_2$O bind to Au[I]? The aqueous solution chemistry of gold is a neglected area and there are very few X-ray studies of gold compounds crystallised from aqueous solution.

Because we are seeking chemical explanations, the discussion will proceed by examining the involvement of gold in its various oxidation states, in turn. In this way we shall not only emphasize the variability of this primary valency, but also of the secondary valency (coordination number). Both will play key roles in events at the molecular level.

Even after 4000 years of study we shall find that there is much scope for future progress in the biological chemistry of gold.

1.1. Gold Drugs — an Early Start

The biological use of gold can be traced back (*1, 2*) as far as the Chinese in 2500 B.C. The element is one of the easiest to obtain in pure form, and up to the eighth century metallic gold was considered to be the cure-all for every known disease. Chemical knowledge was advancing, and by the thirteenth century *Roger Bacon* was recommending Auric chloride (prepared from metallic Au and aqua regia, followed by neutralisation with chalk) for the treatment of leprosy. Even in the eighteenth and nineteenth centuries gold compounds were popular drugs for every disease, always producing favourable results.

Koch's experiments (3) on the effect of AuCN on bacterial growth in 1890 represent the beginning of gold molecular pharmacology and attempts to design gold drugs. At 1 part in 2×10^6 Au(I)CN was lethal to tubercle bacilli in the test tube. However it was much less effective when introduced into the blood serum of infected animals. Attempts to put this discovery to clinical use met with partial success for skin tuberculosis and syphilis, but toxic side reactions were severe. From about 1913—1927 was a period of intense searching for Au(I) compounds of lower toxicity. At this stage organic thiols were introduced as ligands. Table 1 shows some of these compounds.

Table 1. Some gold drugs

Formula	Ligand	Trade Name
Au_2S_3(colloidal)	Sulphide	Aurol Sulphide
$Na_3Au(S_2O_3)_2$	Thiosulphate	Sanocrysin, Crisalbine, Aurothion
[benzene ring: CO_2Na, SAu, NH_2]	4-Amino-2-mercapto-benzoic acid	Krysolgan
[benzene ring: SO_3Na, SAu, $NHCH_2SO_2Na$]	4-Aminomethylsulphinic-acid-2-mercaptobenzene-1-sulphonic acid	Solganal
$CH_2S\,Au$ $CH\,OH$ CH_2SO_3Na	Thiopropanol Sulphonate	Allocrysine
[sugar ring: CH_2OH, O, SAu, OH, HO, H, OH]	β-D-Thioglucose	Solganal-B-Oleosum
$Au-S-CH-CO_2Na$ $\quad\quad CH_2CO_2Na$	Thiomalate	Myocrisin

However, with echoes of the eighth century, there were no controlled trials. Emphasis was simply placed on introducing gold into the system, adjusting doses according to the compound to give the same amount of metal. New compounds were readily acclaimed: "where colloidal gold has failed, aurothiopropanol succeeds"(4).

Gold again caught the public imagination from 1925–1935. These years have been termed the "gold decade" in tuberculosis treatment (5). However laboratory groundwork on the curative effect was insecure, and toxicity still a problem. Although the clinical benefits were eratic, there was an astonishing acceptance of the drug during these years, followed by a rapid rejection without the immediate introduction of a substitute.

It was during the gold decade, in 1927, when *Landé* (6) introduced Au(I) compounds for the treatment of arthritis. He mistakenly assumed that a relationship exists between chronic polyarthritis and tuberculosis. However good results were reported and these compounds have remained in clinical use ever since. According to medical opinion now (7) they are as effective as any other available drug for difficult cases, and are amongst the few drugs that can alter the course of the disease.

From a consideration of the oxidation state diagram we shall see why only a restricted range of gold compounds have emerged as drugs.

2. Oxidation State Diagram

We see in Table 2 that gold forms a bridge in the Periodic Table between the transition metals and the B metals, which have contrasting chemical behaviours (8). With their unfilled d shells of electrons, the transition metals exhibit a wide range of magnetic and spectroscopic properties. The d shell is closed in B metals, and a pronounced tendency toward covalent rather than ionic metal-ligand bonds is apparent.

Thus Au(I) (5 d^{10}) would be expected to resemble the B metal ions, and indeed it forms stronger complexes with the more polarisable anions. The following trends of increasing thermodynamic stability of AuL_2 complexes has been found (9, 10).

Anionic complexes:

$$CNO^- < CNS^- \sim Cl^- < Br^- < I^- \ll CN^-$$

Cationic complexes:

$$Ph_3PO < Me_2S < Py < AsPh_3 < NH_3 < C_6H_{11}NH_2 \ll PPh_3$$

The most stable complexes being those of the π-acid ligands cyanide and phosphine. Au(I) can form weak (π) complexes with olefins, and in this resembles transition metal ions. A general feature of Au(I) compounds is their instability in water. This can readily be seen from the oxidation state diagram, Figure 1. Although the halide complexes $Au(Hal)_2^-$ are stable in non-aqueous media, Table 3, in water they rapidly disproportionate with deposition of metallic Au:

$$3 \ Au(I) \rightarrow 2 \ Au(0) + Au(III)$$

The ligands which stabilise Au(I) in water include CN^- and $S(RS^-, R_2S, S_2O_3^=)$.

Table 2. Gold: Electronic Configuration and Size.

Group			
VIII	IB	IIB	
Ni	Cu	Zn	^{197}Au (100%), atomic number 79
Pd	Ag	Cd	nuclear spin I = 3/2
Pt	*Au*	Hg	
Electronic configuration:			(Xe) $5 \ s^2 \ 5 \ p^6 \ 5 \ d^{10} \ 6 \ s^1$
Isoelectronic ions:			Au(I) − Hg(II), Au(III) − Pt(II)
Radio-isotopes:			^{198}Au, half-life 2.7 days (β^-, → ^{198}Hg)
			^{199}Au, 3.15 days (β^-); ^{195}Au, 183 days (γ)
Approximate radii:			Au(I), 0.137 nm; Au(III), 0.085 nm.

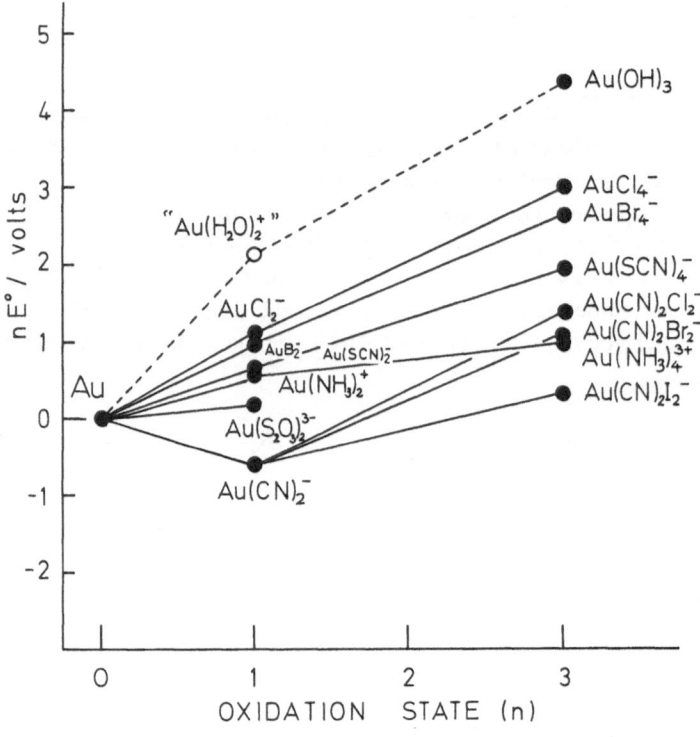

Fig. 1. An *oxidation state diagram* for gold. Points for Au(III) lie high on the graph, indicative of their oxidising properties. Points for many Au(I) complexes lie above a line joining Aû to the relevant Au(III) complex, and the Au(I) complex is therefore thermodynamically unstable with respect to disproportionation into Aû and Au(III). Note the stabilisation of Au(I) by $S_2O_3^{2-}$ and CN^-. The E° for Au(H_2O)$_2^+$ is a calculated value [11]. Data from (*12, 13, 14, 15, 16*).

Table 3. Stability of Au^+ in acetonitrile ($\mu = 0.1$, tetra-ethylammonium perchlorate, 24 °C) (*9, 10*).

Species	log k_1	log k_2	log $\beta_{1,2}$
AuCl, $AuCl_2^-$	12.15	7.79	19.94
AuBr, $AuBr_2^-$	11.98	8.41	20.39
AuI, AuI_2^-	17.1	6.7	23.8
$Au(CN)_2^-$	–	–	33.7
Au(OCN), $Au(OCN)_2^-$	9.77	8.11	17.88
$Au(SCN)_2^-$	–	–	20.2
$Au(pyridine)_2^+$	–	–	11.0
$Au(NH_3)^+$, $Au(NH_3)_2^+$	10.14	8.0	18.14

(AuF, AuF_2^-, $AuCN$, $AuSCN$, $Au(en)^+$ and $AuOH$ are insoluble)

176

Higher oxidation states have unfilled d shells, but Au(II) ($5\,d^9$) and Au(V) ($5\,d^6$), although known, are extremely unstable. Au(II) could play an important part in biological gold chemistry as an intermediate in redox reactions of Au(I) or Au(III), and we shall return to it again briefly.

The points for Au(III) lie high on the oxidation state diagram indicative of the strong oxidising power of complex ions such as $AuCl_4^-$. We note here that biological systems tend to be mildly reducing, operating at potentials around $-0.5\,v$ to $0\,v$.

The diagram illustrates the increased stabilisation of Au(III) by N ligands such as NH_3.

We begin with a discussion of Au(0) and include colloidal gold under this heading.

3. Au^0

3.1. Natural Uptake

Metallic gold, Au(0) is an exceptionally stable form of the element and most deposits occur in this form. Uptake of gold by living things would therefore involve attack on the metal itself. Such a mechanism does not seem to have been widely developed, and, as fas as we know, there is no natural requirement for gold. Certainly there is no absorption of gold through the skin as a result of wearing rings (*17*), although rare skin allergies can occur (*18*).

The natural abundance of gold is low (about 5 mg/ton earth's crust, lmg/ton sea water), but a few plants living in gold-rich soils are able to accumulate gold. *Artemisia persia*, for example, develops high gold concentrations during the first part of its growth period (*19*), whereas *prangos pobularia* has high concentrations during flowering. *Stripa* grasses can contain as much as 100 g of gold per ton (*20*), perhaps a good model for study of the uptake and storage of gold. Microorganisms in the plant roots may be responsible for solubilising gold ready for uptake. It is known (*21*) that some strains of the micro-organism *Bacillus megaterium* secrete amino acids aspartic acid, histidine, serine, alanine and glycine to aid gold dissolution.

3.2. Colloidal Gold

When gold compounds are reduced to metallic gold, colloidal solutions are frequently obtained, i.e. gold particles dispersed in a homogeneous medium. *Zsigmondy* (*22*) has elaborated methods for obtaining gold sols of varying degrees of dispersion. For example, reduction of $HAuCl_4$, neutralised with K_2CO_3, by white phosphorus in ether, gives particle sizes of 1 to 3 nm. *Frens* (*23*) uses citrate as a reducing agent, and can carefully control particle sizes between 16 and 147 nm in diameter. Gold sols are usually very stable[1]), and depending on the reaction conditions, have colours ranging from red, violet ("Purple of Cassius") to blue. The absorption maximum shifts to longer wavelengths as the particles grow larger. Thus, red sols are highly disperse and absorb at about 500–550 nm, whereas in blue sols the particles are larger and the absorption shifts to 580–650 nm. Flocculation of colloidal gold sols is accompanied by a colour change from red to blue and is enhanced by the presence of electrolytes such as NaCl (and those present in serum), but can often be prevented by addition of ethylene glycol, gelatin or albumin.

Because colloidal gold particles are easily seen by electron microscopy, they have been widely used for biological labelling purposes, but can *specificity* of labelling be

[1]) The Royal Institution in London still has 3 blue/ruby-coloured colloidal gold solutions prepared by Faraday in 1856.

expected? To answer this we must ask about the interactions involved in the labelling process. Two illustrations will now be given.

(i) Uptake of Colloidal gold by macrophages

Macrophages are cells which "drink" or "eat"[2]) material from the surrounding medium. This process removes "foreign bodies". The mechanism appears to involve the formation of tiny invaginations of the surface membrane to give vesicles which move toward the centrosomal region of the cell. Here, these pinocytic vesicles fuse together to form vacuolar structures 500—1000 nm in diameter. This is of special interest in active macrophages because the pinocytic vacuoles soon acquire hydrolases and become lysosomes.[3])

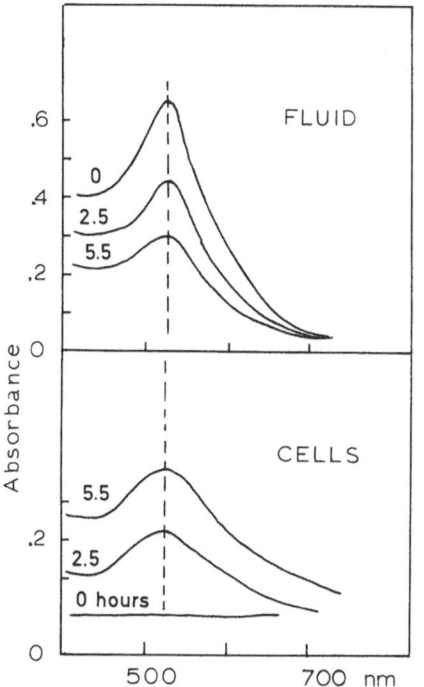

Fig. 2. The upper traces illustrate the disappearance of the colloidal Au from solution after incubation for t (hours) in the presence of macrophages, measured as a decrease in optical density at 525 nm. The lower traces show the increasing intensity of colour of the macrophages separated from the above solution. (From (24)).

---

[2]) *pinocytosis* (drinking) is the uptake of fluid droplets by cells, and a cell can typically increase its volume by 1/3 in 1 hour.
phagocytosis (eating) is the uptake of particles by cells. The two terms are often used interchangeably.
[3]) A *lysosome* is a bag of hydrolytic enzymes used for intra- and extracellular digestion.

Colloidal gold is one of many materials which can be taken up by macrophages. *How does the macrophage recognise a gold particle: by size?, surface charge?* Gosselin proposed (*24*) that uptake occurs in two stages: a reversible absorption of gold onto the cell surface, followed by an irreversible ingestion process. He followed the uptake spectrophotometrically at 525 nm, see Fig. 2. The red colloid became blue when ingested by the cells, indicating a coagulation of the gold. Less than 1 % of the surface has bound gold, and of this 2–20 % was ingested per minute. Once taken up, the gold remains in the cells for at least 24 hours (*25*). More insight into the mechanism of uptake comes from studies of cell sections by electron microscopy (*26*). It appears that gold particles are taken into the cell in small pinocytic vesicles, which form by invagination of the cell surface membrane, and these vesicles fuse with one another to form vacuoles while being transported to the centrosomal region. A typical pinocytic vacuole containing many gold particles is shown in Fig. 3.

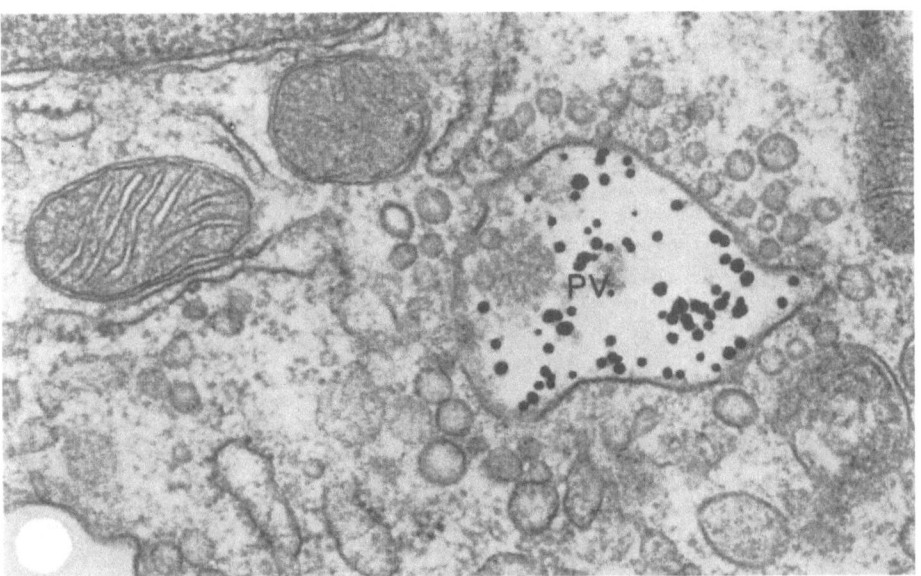

Fig. 3. A pinocytic vacuole (PV) in a *macrophage* which had been cultured for several hours in a medium containing colloidal Au. A part of the nucleus (upper left) and mitochondria (M) are also visible. The pinocytic vacuole is filled with particles of colloidal Au, but the few vesicular structures inside and the numerous vesicles outside do not contain Au, which indicate that most of them are probably not pinocytic vesicles arising from cell surface or by budding from the pinocytic vacuole (× 60000) (*26*).

The part played by the surface of the gold particle in triggering the chemical message which induces phagocytosis is unknown, but perhaps the message is based on a reaction of Au(I) ions on the surface. The surfaces of colloidal Au particles are known to become charged through absorption of ions (22), and Pauli has suggested (27) that $AuCl_2^-$ ions are responsible for some of this charge.

There is evidence (28) that a soluble gold-uptake-stimulating factor of molecular weight $< 100\,000$ is secreted by lymphocytes and acts upon macrophages. This factor may be a conventional antibody, and may interact specifically with the surface of gold particles. There is already some evidence that immunoglobulins (antibodies) can bind to colloidal gold, which we shall now describe.

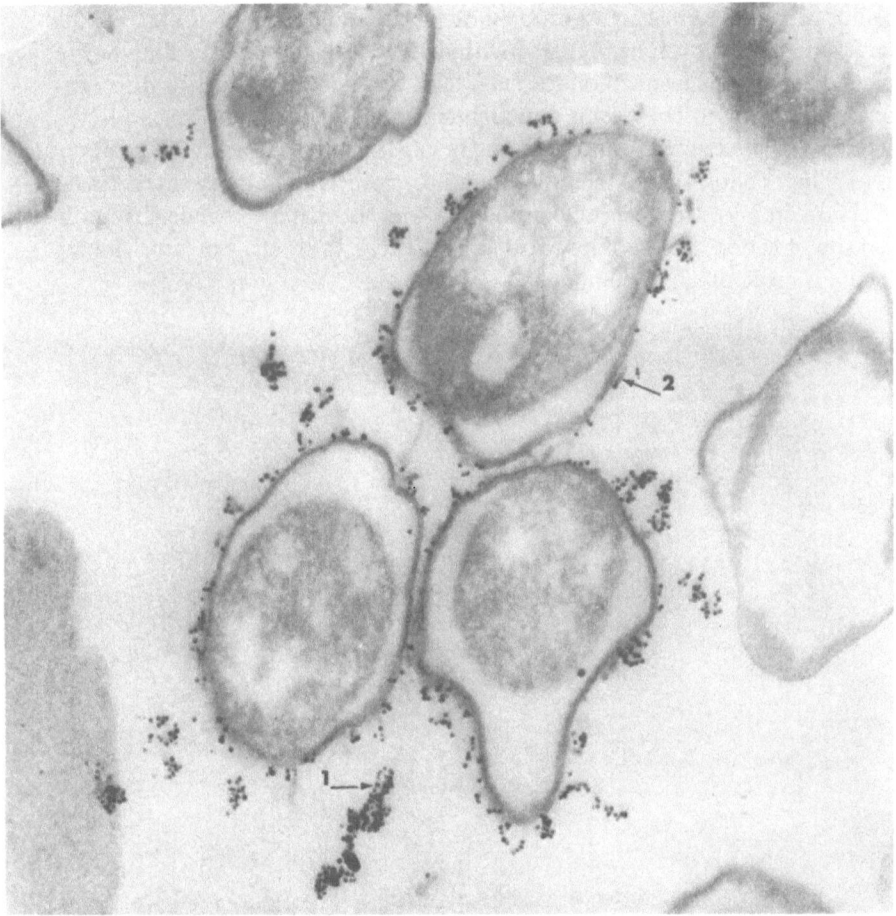

Fig. 4. *Salmonellae bacteria* coated with rabbit ant-salmonella anti-bodies (gamma-globulins) labelled with colloidal Au (black clots). Seemingly unattached immunocolloid (arrow 1) are attached to free pieces of cell wall (arrow 2), magnification × 125000. From (29).

(ii) Labelling of immunoglobulins and proteins

Proteins such as serum albumin have long been used to prevent flocculation of colloidal gold sols: there must be an interaction between them. *Faulk* and *Talyor* (*29*) have labelled antibodies with colloidal gold to probe cell-surface binding sites. They found that rabbit anti-salmonellae serum adsorbed onto particles of colloidal Au retained full activity, and when the immunocolloid was exposed to salmonellae bacteria, the distribution on the cell surface was clearly visible by electron microscopy, Fig. 4.

The interactions involved may be electrostatic rather than covalent in nature, i.e. colloid surface-AuX_2^- ... $H_3\overset{+}{N}$—(Lys)—antibody, rather than colloid surface —Au—S—(Cys)—antibody, for, although horse anti-human IgG can be labelled with an average of 1.6 particles (diameter 3 nm) of Au per antibody molecule, human or rabbit IgG cannot be labelled on account of rapid dissociation on adding NaCl (*30*).

Since gold-particles can be prepared with defined sizes, double labelling techniques are possible (*31*). Thus, mannan on the surface of *Candida utilis* is easily visible by scanning electron microscopy using anti-mannan antibodies or Concanavalin A.

Attempts to use colloidal gold as a drug in experimental animals have been unsuccessful (*32*), but colloidal [198]Au has been used for destruction of unwanted tissue (e.g. in synovectomy). However, one problem is that the material wanders from the required site of action, and perhaps another is its decay to [198]Hg. For drugs more specific use is made of Au(I) compounds.

4. Au(I): Aurous Compounds

Controlled clinical trials by the Empire Rheumatism Council (*33*) have demonstrated that aurothiomalate and aurothioglucose can be effective for the treatment of rheumatoid arthritis, although the disease is now known to be unrelated to tuberculosis which is caused by a *mycobacterium*. For the *difficult cases* aurothiomalate or aurothioglucose are as effective as any other available drug (*7*). Around 50% of patients receiving these are usually markedly improved or cured (with careful choice of patients and administration conditions the figure may be higher; *Dr. Gumpel*, personal communication), but toxic side effects are observed in about 40% of cases. These are of such concern that a leader (*34*) in a medical journal recently asked "Is gold therapy worth the risk?" We now explore a possible molecular basis for these therapeutic and toxic reactions.

4.1. Arresting the Growth of Microorganisms

$Au(CN)_2^-$ is lethal at 1 part in 2×10^6 to *tubercle bacilli* in a test tube. However, *Koch* (*3*) in 1890 was unable to demonstrate an *in vivo* effect. Some cultures of the Koch bacillus can develop resistance to Au^I salts such as allochrysin, and grow in media containing much higher gold concentrations (*35*). Such resistance to heavy metal ions may be due to the presence of extra-chromosomal plasmid DNA, such as that which has been demonstrated in *straphylococcal aureus* (*36*). Pathogenic and non-pathogenic classes of straphyloccal are known to differ in their resistance to Hg^{2+} ions: hospital strains of the organism have a larger resistant proportion (*37*).

Au(I) compounds can also effectively control mycroplasma[4]). *Sabin* and *Warren* (*32*) have demonstrated in vitro inhibition of their type B strain, which can cause gold-treatable rodent arthritis. However mutant strains, which are resistant to gold, can remain active even when grown in the presence of Au(I) for three generations (*38*).

Mycobacteria — and mycoplasma-induced arthritis are both used as models for testing possible new gold drugs. Intradermal injections of *Mycobacterium butyricum* in animals are often used. However, the relationship to true arthritis is unknown and the model may be a test of *anti-inflammatory activity alone*.

---

[4]) *Mycoplasma* are a genus of bacteria-containing cells which do not possess a true cell wall, only a thin membrane. They are found in humans and other animals.

4.2. Au(I)-*Sulphur Ligands*

As we have seen the choice of Au(I) thiols as drugs is largely a result of the limited range of ligands that stabilize Au(I) in aqueous solution. Many of these complexes are not indefinitely stable, but often decompose slowly on standing in solution at room temperature for a few days. For example, colourless aqueous solutions of aurous thiosulphate soon become yellow, although the crystal structure of this compound is one of the very few in this class which has been determined (*39*), see Fig. 5. Most of the other compounds are white or pale yellow solids, prepared, typically, from an Au(III) salt and reducing agent, (SO_2 or excess thiol) plus thiol.

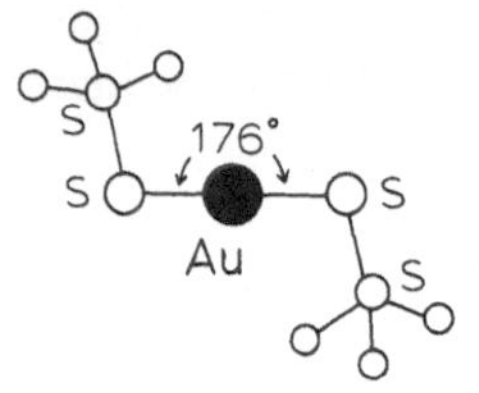

Fig. 5. The $Au(S_2O_3)_2^{2-}$ anion in crystalline $Na_3Au(S_2O_3)_2 \cdot 2H_2O$ (Sanocrysin). The Au–S bond distance is 0.228 nm, and the nearest neighbour to the gold atom, besides S is another gold atom at 0.330 nm. The water molecules are H-bonded to S and O atoms of thiosulphate ligands (*39*).

(i) Clinical Use

Disodium aurothiomalate (Myocrisin) and aurothioglucose (Solganal) are in clinical use today for the treatment of rheumatoid arthritis. Patients receive 50 mg of aurothiomalate in 0.5 ml H_2O (intramuscular injection) every week until about 1g has been administered. About 30% of each dose seems to be excreted within 1 week. Commonly a deterioration occurs at 0–3 months after treatment commences, with improvements at 3–4 months. Serum Au levels are kept high (about 300 μg Au per 100 ml serum), and although side-effects occur in about 40% of adults (*Dr. B. M. Ansell*, personal communication, e.g. in bone marrow, kidneys or skin) patients have been maintained on gold treatment for up to 20 years. Eventually, gold seems to enter almost every cell in the body, and is removed at a very slow rate. A level of 22 μg Au/100 ml serum has been found in the serum of one patient 9 years after treatment had ceased (*Dr. R. Grahame*, personal communication).

(ii) Mechanisms of Action

There are many speculations about the mechanism of action of gold as anti-arthritic drugs Table 4. We examine some of these below.

Table 4. Summary of some possible mechanisms of action of $Au^I S$ anti-arthritic drugs.

1. Inhibition of lysosomal enzymes in phagocytic cells which are responsible for causing joint inflammation.
2. Arrests growth of the infectious agent (bacteria, mycoplasma, virus?) responsible for the disease.
3. Affects joint structure through stabilisation of collagen by cross-links or altering abnormal disulphide bonding in proteins.
4. Inactivates granulocyte elastase, and inhibits other granulocyte enzymes (toxic action).
5. Affects immune response.
6. Affects prostaglandin synthesis, and therefore inflammation.

At the joint – in lysosomes of macrophages

Gold drugs may act at the joint itself. As can be seen in Fig. 6. Au accumulates in the synovium during gold treatment ("chrysotherapy"). The distribution is not uniform, but the gold is concentrated in packets within macrophages[5]) (*40*), those cells which we have already seen to ingest colloidal Au. If we take a closer look at an individual macrophage, we find that the gold is further localised, within the lysosome, Fig. 7. This is such a distinctive feature that lysosomes loaded with Au are often termed "aurosomes". Since joint damage in rheumatoid arthritis could be caused by the release of hydrolytic lysosomal enzymes within the inflamed joint, it is appropriate to ask whether Au(I) thiols cause inhibition of these enzymes. *Persellin* and *Ziff* (*43*) found that the activities of acid phosphatase and β-glucuronidase are reduced after incubating guinea pig peritoneal macrophage extracts with aurothiomalate. Some inhibition of the mitochondrial enzyme malate dehydrogenase was also observed. This enzyme has a Cys sulphydryl group at its active site. Is the thiomalate ligand released when aurothiomalate binds?

4.3. Au(I) *on Proteins and Enzymes*

There have been very few detailed studies of Au(I) binding sites, and most of these involve $Au(CN)_2^-$, Table 5. Liver alcohol dehydrogenase binds $Au(CN)_2^-$ ($K = 1.6$ mM^{-1}) which competes with the phosphate binding site of NAD$^+$ on the enzyme (*44, 45*). The cyanide ligands appear to remain bound to gold, despite the presence of two very

---

[5]) Using the skin-window technique it has been shown that aurothiomalate administration inhibits the *in vivo* phagocytic activity of inflammatory macrophages (*41, 42*).

185

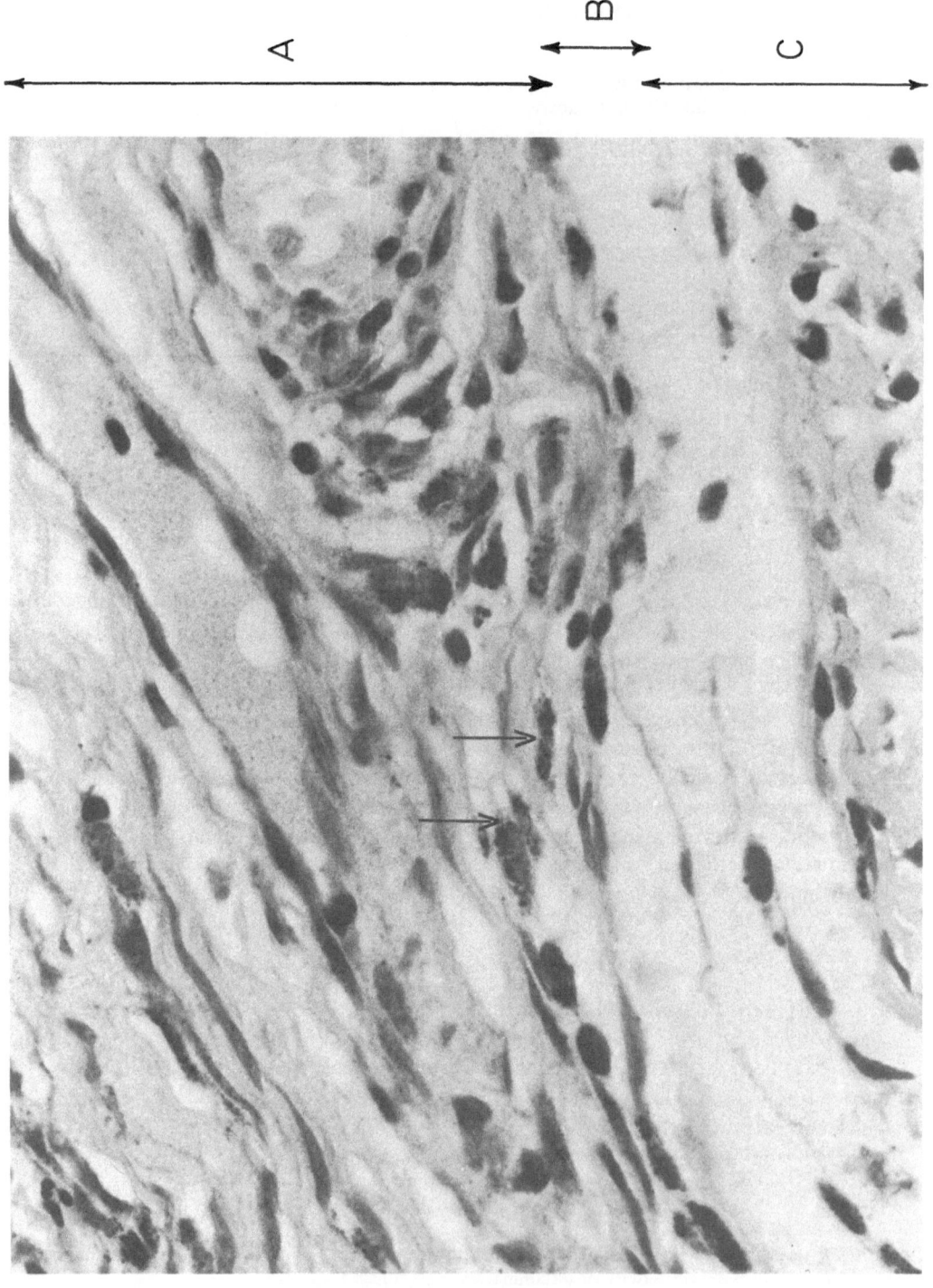

reactive Cys SH groups, on the enzyme. Hg^{2+}, on the other hand binds strongly to these groups, and denatures the enzyme. Carbonic anhydrase also has a free sulphydryl group, but $Au(CN)_2^-$ binds by bridging His 128 and a zinc bound water molecule (46). This is illustrated in Fig. 8. It would not have been surprising if the Au(I) ion had increased its coordination number from two to three or four when $Au(CN)_2^-$ became bound to the enzyme. Indeed, we know that Au(I) can adopt a coordination number greater than two, see Table 6. Two enzyme binding groups (such as histidine) could possibly mimic the behaviour of $\alpha\alpha^1$-dipyrridyl or o-phenanthroline which form

Table 5. $Au(CN)_2^-$ labelling of enzymes

Enzyme	Conditions	Au Site	Reference
Liver Alcohol Dehydrogenase	1 mM $Au(CN)_2^-$, Tris buffer, pH 8.4, pentanediol	2 sites, competes with NAD^+ phosphates	44,45
Carbonic Anhydrase B	1 mM $Au(CN)_2^-$ pH 8.2, 2.3 M $(NH_4)_2SO_4$	2 sites 45%/14% occupancy	47
Carbonic Anhydrase C	20 mM $Au(CN)_2^-$ pH 8.5, 2.3 M $(NH_4)_2SO_4$	linear $Au(CN)_2^-$ linking Zn bound H_2O and His 128.	46

Table 6. Some Variations of Au(I) coordination number.

Example	gold coordination number	geometry
$Au(CN)_2^-$		
AuCN	2	linear
$AuCl(PCl_3)$		
$AuCl(PPh_3)_2$	3	trigonal planar
$Au(CN)_2$ (o-phen)	4	square-planar
Au(ditertiaryarsine or phosphine)$_2^+$	4	tetrahedral
Linear Au^I complexes with Au−Au bonds included	4	square-planar

◁ Fig. 6. Section of *human synovium* (about × 200) from a patient with rheumatoid arthritis currently receiving aurothiomalate, showing greatly inflated subintima (area A) containing many cells laden with gold grains, separated from the deeper joint tissue (area C) by a layer of fatty tissue area B. Two examples of macrophages containing granular deposits of reaction product following staining for gold are arrowed. (*B. Vernon-Roberts* and *J. L. Doré*, personal communication).

187

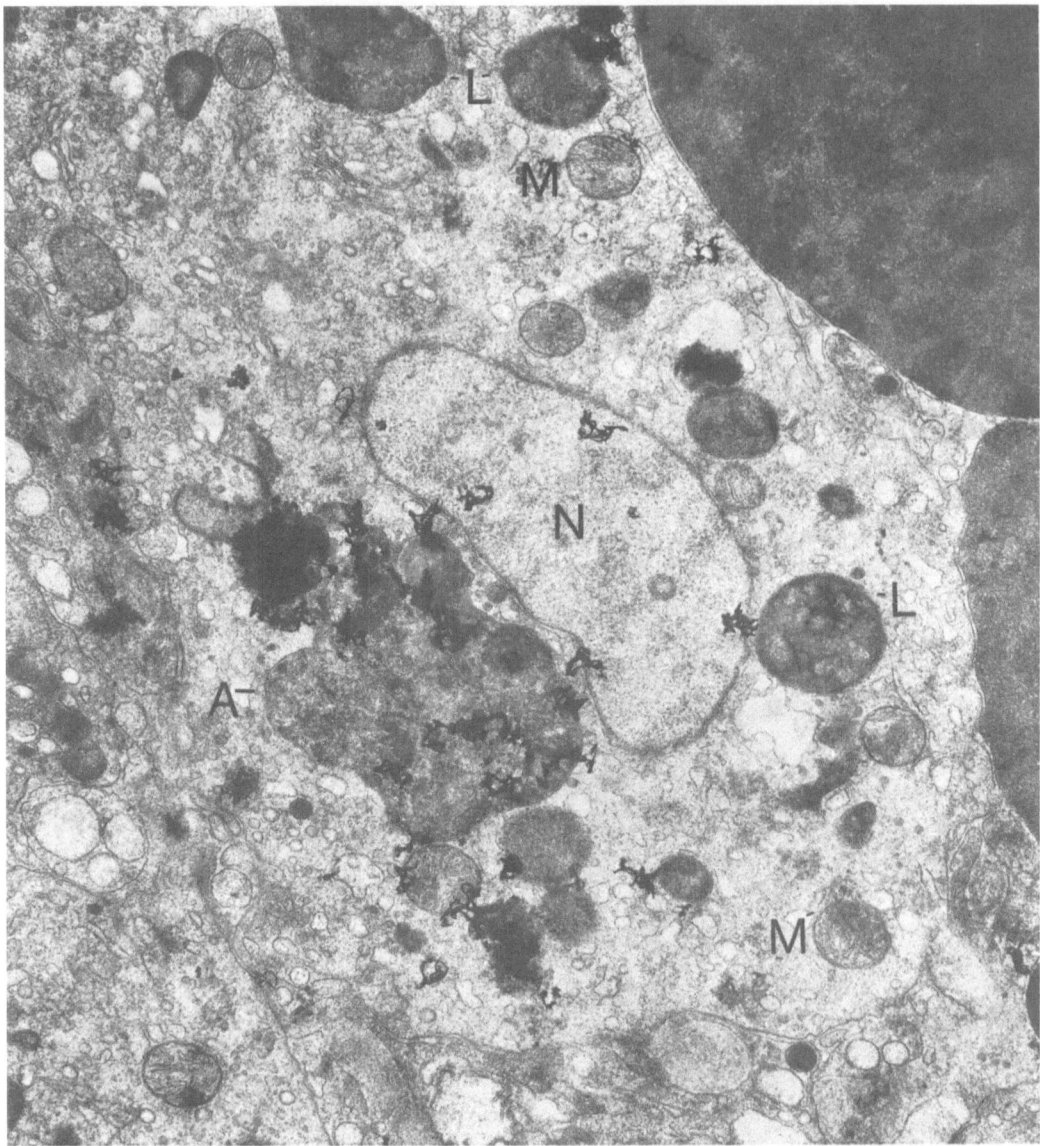

Fig. 7. A rabbit spleen *macrophage* (magnification x 22500), fixed with osmium tetroxide, following injection of ^{198}Au aurothiomalate. The black squiggles represent the decay of ^{198}Au which can be seen to be localised on the membrane of the nucleus (N), membrane of mitochondria (M) and within lysosomes (L). A lysosome in the centre of the picture has a particularly high concentration of ^{198}Au and is termed an *aurosome* (A). Lysosomes are bags of hydrolytic enzymes used for the orderly destruction of cellular components. (*B. Vernon-Roberts* and *J. L. Doré*, personal communication).

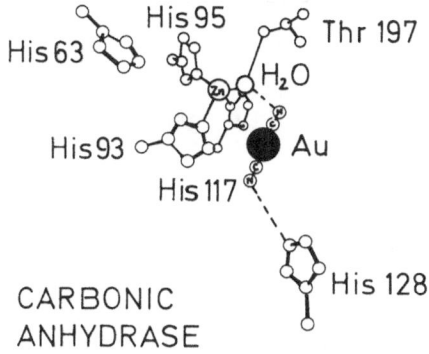

CARBONIC
ANHYDRASE

Fig. 8. The Au(CN)$_2^-$ binding site in human carbonic anhydrase C, as determined by X-ray diffraction. Several H$_2$O molecules are displaced when the anion binds. One cyanide is hydrogen bonded to the Zn-coordinated H$_2$O molecule, the other is within hydrogen bonding distance of His 128 (46).

square – planar addition complexes, e.g. Au(CN)$_2$ (dipy)$^-$ (48). However we note that Au(CN)$_2^-$ does not readily take up an additional CN$^-$ to form Au(CN)$_3^=$ (49).

Aurothiomalate is reported to inhibit enzymes involved in the synthesis of uridine-5'-phosphate in granulocytes, although the drug is not taken up by intact granulocytes (50, 51). Human granulocute elastase also shows a 40–60% inhibition (52).

Bis(thioacetoxy)Au^I is a convenient histochemical substrate for the enzyme acetylcholinesterase (53). Black, insoluble deposits of Au$_2$S are one of the products and serve to pin-point the localisation of this enzyme by electron microscopy.

4.4. Albumin and Immunoglobulin Complexes

About 85% of the gold in circulation after injection of aurothiomalate is bound to serum albumin (54, 55), a protein of molecular weight 65000. There appears to be about half as much gold in the synovial fluid as in serum, and gold disappears from both of them with similar rates, half-life about 5.5 days (56). Is the gold bound to the sulphydryl group of this protein? Gerber (57) found that aurothiomalate altered the heat and urea denaturation of serum albumin, and suggested Cys SH binding was involved, thereby preventing the formation of inter-protein disulphide bonds. He noted that no free thiomalate was released into solution. However it is difficult to remove the gold from albumin using competitive SH reagents such as cysteine, mercaptoethanol or p-hydroxymercuribenzoate (58). The structure of albumin is unknown, but a Cys SH group is thought to be located in a crevice 0.95 nm deep (59).

Some of the aurothiomalate is presumably precipitated and dispersed at a very slow rate, since the Ca^{2+} and Mg^{2+} salts are insoluble and serum contains about 3 mM Ca^{2+} and 1 mM Mg^{2+}.

At high serum Au levels (e.g. 300–700 μg per 100 ml) there is significant binding to other serum proteins such as immunoglobulins (45), as shown in Fig. 9. These gold immune complexes may be taken up by phagocytic cells (60). Aurothiomalate inhibits the Cu^{2+}-catalysed aggregation of human gamma globulin (61).

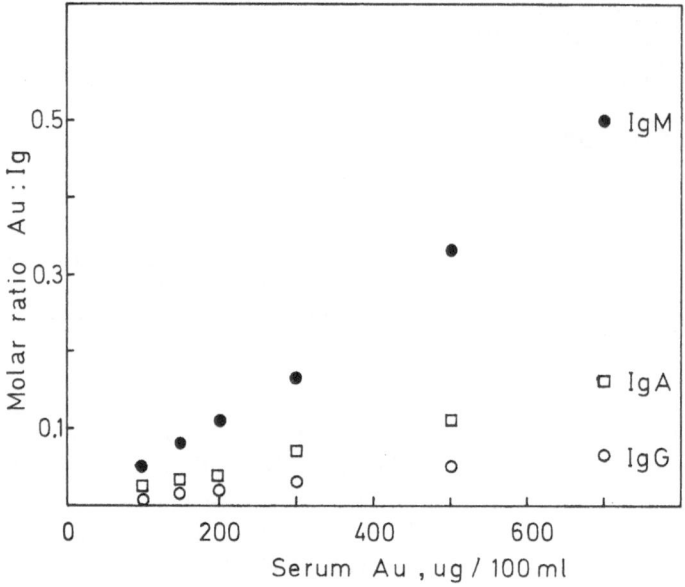

Fig. 9. Graph illustrating the increased affinity of gold for various immunoglobulins (Ig) as total serum gold concentration increases. Molecular weights: IgG – 160000, IgA – 390000, IgM – 1000000. Data from (*54*).

The effect of Au^I drugs on the immune response is not clear. *Denman* and *Denman* (*62*) found that aurothiomalate was able to induce lymphocyte transformation in those patients who became sensitised to the drug, and patients with haematological abnormalities give positive lymphocyte transformation tests on separose beads labelled with aurothiomalate (*G. Stafford* and *A. M. Denman*, personal communication).

Does Au^I-binding interfere with the ability of albumin to transport Cu? Since Cu is often employed as a Cu^I/Cu^{II} redox couple (e.g. in cytochrome oxidase); Au^I would not be expected to be an active replacement. Au^I is also a much larger ion (by 0.04 nm in radius).

An understanding of the binding of Au^I thiols to proteins or enzymes requires a knowledge of the nature of the interacting gold complex.

4.5. Molecular Structures of Au^I Thiols

In the solid state and in solution we must expect Au(I) to adopt a coordination number of 2, 3, or 4, see Table 6. Two-coordination is very common, three coordination may be common in solution because of the ease of ligand exchange on Au^I (intermediates, *vide infra*), and four coordination including Au–Au bonding, may be more common

190

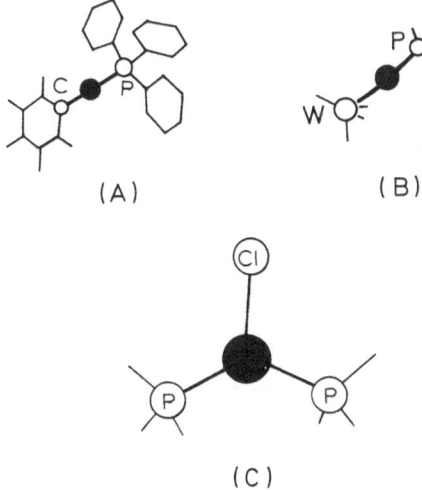

(A)

(B)

(C)

Fig. 10. (*A*) The linear coordination (178°) about Au^I in (pentafluorophenyl) (triphenyl-phosphine) Au^I. There are no intermolecular contacts with the Au atom of less than 0.35 nm (*63*). (*B*) Linear coordination in $(\pi-C_5H_5)W(CO)_3AuPPH_3$ (*64*). (*C*) Chloro-bis(triphenylphosphine) Au^I has trigonal planar coordination about Au^I. The compound is formed when ClAuPPh₃ is refluxed with excess PPh₃ in benzene. (*65*).

than presently recognised. There is, however, a paucity of data, particularly on x-ray studies of compounds crystallised from aqueous solution. Fig. 10 shows 2- and 3-co-ordinate Au(I) compounds.

Recent work (*66*) confirms an earlier suggestion (*67*) that aurothiomalate exists as a polymer in aqueous solution. The degree of association is now known to be dependent on the ionic strength of the solution. For a 10^{-2} M solution in 0.5 M NaCl, the average degree of polymerisation is about 6. Specific broadening of the ^{1}H NMR spectrum of aurothiomalate in the presence of NaCl is observed (*66*), and a similar result is obtained with NaClO₄ and other salts. Both ^{1}H and ^{13}C NMR indicate a highly specific inter-molecular association of aurothiomalate, Fig. 11. This can also be demonstrated for aurothioglucose, (*A. A. Isab* and *P. J. Sadler*, unpublished), and can occur in serum.

Does aurothiomalate polymerise and provide linear 2-fold coordination for each Au^I ion? This is the case for AuCN and AuCl (*68, 69*) which form chains with bidentate cyanide or bridging Cl ligands:

e.g. ... $-Au-C-N-Au-C-N-Au-$

Some other examples of linear coordination are given in Fig. 4 and 10. In view of the lack of affinity of Au^I for oxygen ligands it seems unlikely that the COO^- groups of the thiomalate ligand will bind to gold either intra- (cannot achieve linear coordina-tion in this way) or intermolecularly. Bridging by sulphur is more likely. This is already known for some Ag^I thiols. For example, there are infinite Ag–S frameworks in Ag^I cyclohexanethiolate (*70*), giving 2- and 3- coordinate Ag^I ions, see Fig. 12. *Hesse* and *Nilson* (*71*) have summarised some of the arrangements of metal ions (A) and coordi-nated ligand atoms (X) commonly found in AX polymers, Fig. 13. It is possible that Au–Au bonding contributes to the stability of Au(I) thiols.

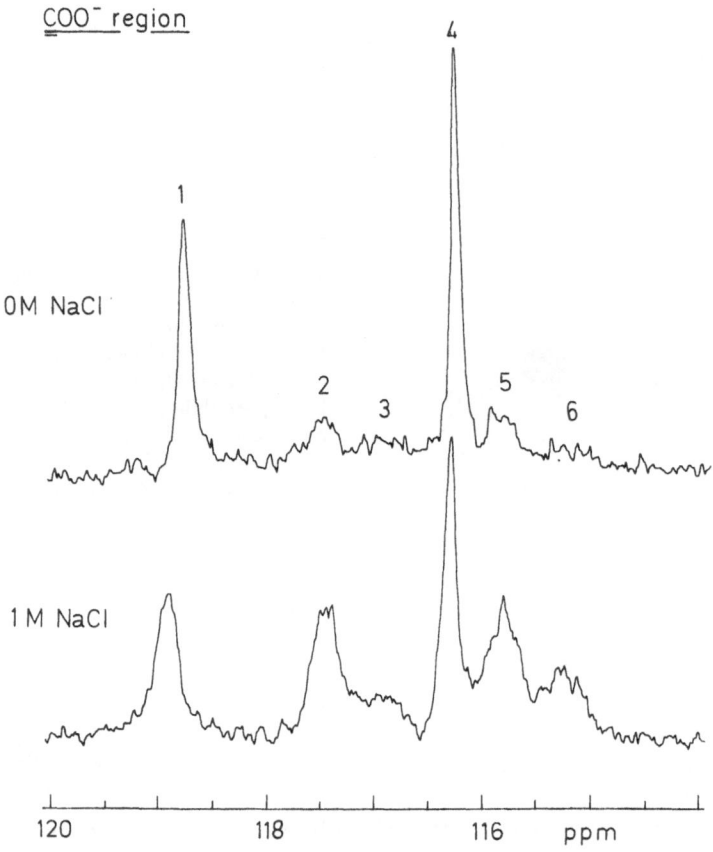

Fig. 11. [^{1}H] noise-decoupled ^{13}C NMR spectra of aurothiomalate (0.8M, in D$_2$O, pH* 7). The broad peaks 2, 3, 5 and 6 which are probably due to polymeric forms increase in intensity on addition of NaCl (*66*).

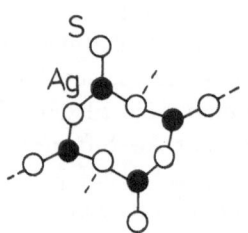

Fig. 12. Representation of the Ag^I coordination in Ag^I cyclohexanethiolate, AgS C$_6$H$_{11}$. The structure consists of *infinite Ag–S frameworks.* Eight of the silver atoms per unit cell have 3-fold almost planar coordination (4 shown) while four of them have two-fold almost linear coordination (involving bonds to S atoms indicated by dotted lines). The Ag–Ag distances within the chain, 0.291 to 0.311 nm, and in the cross-links, 0.303 to 0.327 nm, are slightly longer than those in metallic Ag itself, 0.289 nm (*70*).

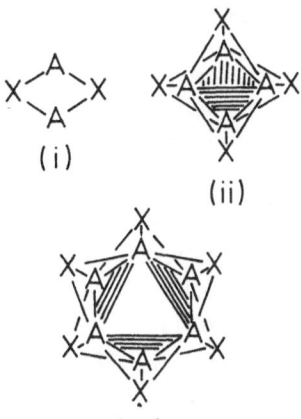

(i)

(ii)

(iii)

Fig. 13. The linkages in some AX polymers, after (71).
(i)   dimer e.g. C_3F_7COOAg, Au^I di-iso-propyldithiocarbamate.
(ii)   tetramer e.g. Cu^I diethyl-dithiocarbamate, $LiCH_3$, Et_3AsCuI
(iii)   hexamer e.g. Ag^I dipropyldithiocarbamate, Ag^I dipropylmonothiocarbamate.

Au^I–Au^I interactions have been postulated in several compounds. The most thoroughly investigated is Au^+–di-isobutyldithiocarbamate, Fig. 14. This compound has a dimeric structure in which the gold atoms are only 0.276 nm apart. This is 0.12 nm shorter than in metallic Au itself, see Table 7. The Raman spectrum of the dimer shows a moderately intense bond at 185 cm^{-1} assignable to Au–Au stretching (81), the bond perhaps being Au^0–Au^0 in character due to charge-transfer from the S ligands onto Au^I. Au–Au bonding seems to be an indispensible force in the stabilisation of crystals of several Au^I compounds, see Figures 15 and 16, with this interaction included the coordination about Au^I becomes approximately square-planar. Many such Au–Au interactions are present in the novel gold cluster compounds shown in Fig. 17 (72, 76), perhaps a pointer to the structure of colloidal gold.

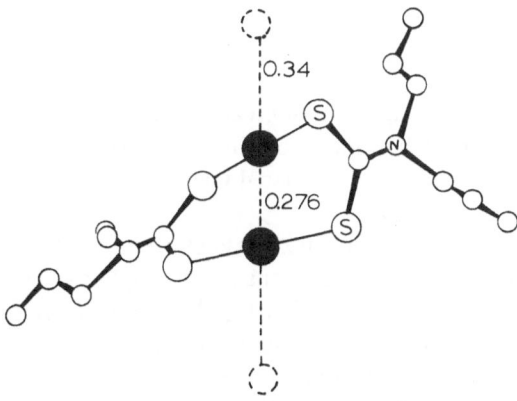

Fig. 14. The dimeric structure of Au^I N,N-n-propyldithiocarbamate. There are alternating short and long Au–Au contacts. The isobutyl analogue has a similar structure (73).

Table 7. Au–Au separations in some gold compounds.

Compound	Au–Au (nm)	Reference
$Au_{11}(SCN)_3[P(C_6H_5)_3]_7$	0.267	72
$[Au^IS_2CN(C_4H_9)_2]_2$,	0.276	73
$[Au^IS_2CN(C_3H_7)_2]_2$		
Metallic Au	*0.2884*	74
$\{[Au(iC_3H_7O)_2PS_2]_2\}_n$	0.29, 0.31	75
$Au_{11}I_3[P(C_6H_4Cl)_3]_7$	0.298	76
$Au^ICl \cdot PCl_3$	0.314	77
$\mu[1,2$-bis(phenylthio)ethane]chloro Au^I	0.319, 0.321	78
Au^{III}(dimethylglyoxime)$_2$ Au^ICl_2	0.326	79
$\{[(iC_3H_7NH_2)Au^I(C\equiv C \cdot C_6H_5)]_2\}n$	0.327	80
$Na_3Au^I(S_2O_3) \cdot 2H_2O$	0.330	39
AuCl	0.336	69

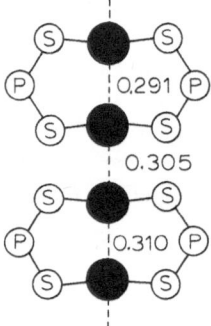

Fig. 15. The gold-sulphur-phosphorus skeleton of two of the dimers in crystalline $\{[Au(i\text{-}C_3H_7O)_2PS_2]_2\}_n$. The *gold-gold contacts extend throughout the crystal*, and each gold atom is in a slightly distorted square-planar configuration (75).

Phosphines are strong Au^I-binding ligands and break up the dimeric units of Au^I dithiocarbamates. Thus the gold atom has linear coordination in (triphenylphosphine) (N,N-diethyldithiocarbamato) Au^I, with a coordinated P and monodentate (through S) dithiocarbamate (82).

The early work on the association of aurothiomalate by freezing point depression (67) concluded that the stability of aqueous solutions of Au^I-sulphur drugs was dependent on the degree of association. Thus $Au^I(S_2O_3)_2^{3-}$ is monomeric (see Fig. 4 which shows that it is monomeric in the crystal) and is very unstable.

It is clear from the NMR measurements that the association of aurothiomalate is highly specific, and even the lowest molecular weight species may be dimers. Tetramers, hexamers and higher order polymers may be present depending on the ionic strength of the solution, with 2- or 3-coordinate Au^I. The association may also be pH dependent.

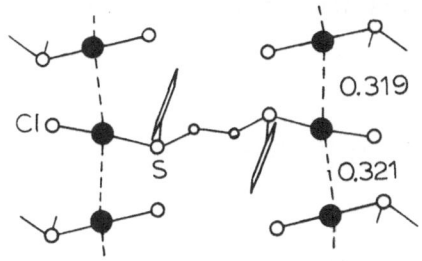

Fig. 16. The crystal structure of μ-[1,2-bis(phenylthio)ethane]-bis[chlorogold(I)]. The Au atoms exhibit linear, 2-fold coordination if only the bonded S and Cl atoms are considered, but this becomes square-planar 4-fold coordination if the nearest-neighbour Au atoms are included. The mode of molecular packing appears to indicate that the polymeric .. Au .. Au .. Au .. chain plays an indispensible part in crystal stabilisation (78).

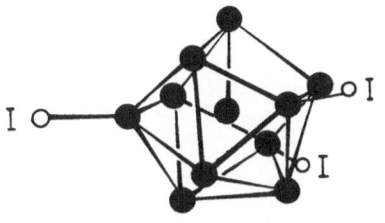

Fig. 17. The $Au_{11}I_3$ group in $Au_{11}I_3$-$[P(C_6H_4-p-Cl)_3]_7$ (76). (P ligands not shown.)

We find that the ^1H NMR spectrum of aurothiomalate in acid solution is almost too broad to be observed (Isab and Sadler, unpublished) which may be a consequence of higher order polymer formation, or a change in exchange rate of the thiol ligand on and off Au^I, due to H^+ competition.

Polymerisation of gold drugs may be a necessary preliminary before they can be taken up by macrophages (see 3.2) if size of particle is an important recognition factor to these cells, although gold uptake via albumin or immunoglobulin complexes is also possible (see 4.4.).

This polymerisation may also explain some of the difficulties in interpreting pharmacological binding data on these drugs (58). An examination of ligand exchange rates on Au^I is also required. Can aurothiomalate be expected to release its thiomalate ligand when it encounters other thiols such as glutathione and cysteine in the biological system?

4.6. Ligand Exchange on Au^I

Ligand exchange rates often determine the reactivity of metal ions. For if ligand exchange is slow, thermodynamically favourable reactions may not occur. Familiar examples of inert metal ions include Co^{III}, Cr^{III} and Pt^{IV}. There is only a limited amount of data of Au^I, but these suggest that exchange is facile.

^{35}S-labelled thiourea exchanges very rapidly with coordinated thiourea molecules in $Au(thiourea)_2^+$ ions (83). In Fig. 18 the rapid exchange of excess phosphine with

CH_3AuPMe_3 is illustrated (*84*). Such processes indicate the ease of formation of 3-coordinate Au^I intermediates:

$$CH_3-Au-PMe_3 \;+\; {}^*PMe_3 \;\to\; CH_3-Au \begin{matrix} {\cdot}^{\cdot}\, PMe_3 \\[4pt] {\cdot}_{\cdot}\, PMe_3 \end{matrix} \;\to\; CH_3Au{}^*PMe_3 \;+\; PMe_3\;.$$

Both Cu^I and Ag^I bisthiourea chloride complexes crystallise with 3-coordinate (perhaps 4-coordinate, if long M–Cl distances are included) metal ions (*85, 86*). They are spiralling polymers with S bridging the metal ions.

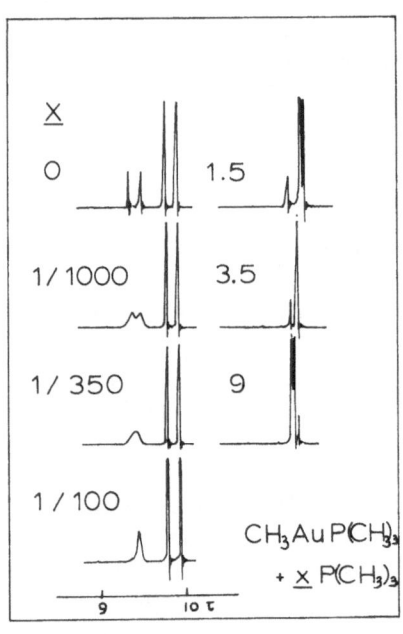

Fig. 18. ^1H NMR spectra of trimethylphosphine methyl Au(I) in benzene solution at various mole ratios of complex and free ligand. The two methyl signals are split into doublets through coupling to ^{31}P. With excess phosphine *exchange of free and bound phosphine* occurs at the CH_3Au centre and the CH_3Au doublet collapses to a sharp singlet ($^1/100$) and shifts upfield. Further addition of phosphine causes a shift of the phosphine methyl resonance towards its unshifted position. $J('HC-{}^{31}P)$ has a different sign in free and complexed phosphine and causes a change in the separation of the doublet peaks. The exchange probably occurs via 3-coordinate Au^I (*84*).

Chadwick and *Sharp* (*49*) however have suggested that exchange of CN^- with $Au(CN)_2^-$ will be slow because of the non-existence of $Au(CN)_3^=$ and by analogy with the isoelectronic $Hg(CN)_2$. There is no experimental data at present to support this.

Hawkins et al. (*11*) have noted that although the normal coordination number of Au(I) is two, complex formation between a phosphine (diphenylphosphinobenzene-m-sulphonate) and $Au(SCN)_2^-$ in aqueous solution produces a *tri*-phosphine Au(I) complex of high stability.

By NMR, a ready exchange of coordinated thiomalate with other thiol ligands such as glutathione can be demonstrated, but not with methionine or histidine (Isab and

196

Aurothiomalate + γ-GLU – CYS – GLY

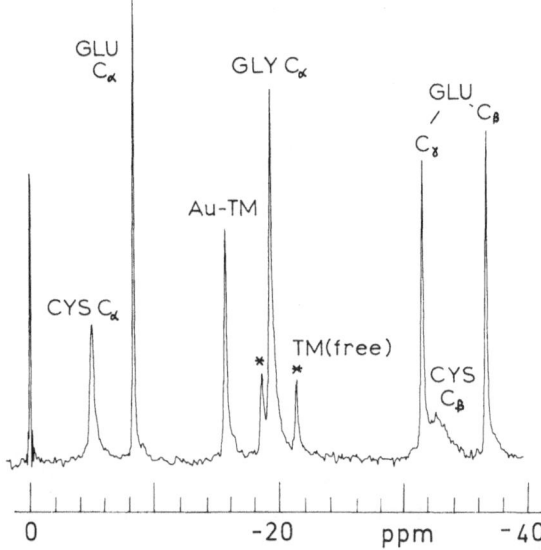

Fig. 19. ^{13}C NMR spectrum ('H-noise-decoupled) of an aurothiomalate-glutathione mixture (1:1, 0.24 M) in water at pH 7. The CH and CH_2 carbon atoms of thiomalate bound to Au have almost the same chemical shift (Au−TM). About half the thiomalate is released free into solution (TM, free). Glutathione is bound to Au by the CYS sulphydryl group as can be seen by the broadening (and shift) of CYS C_α and C_β relative to unbound glutathione. The reference is the CH_2 peak of glycerol. (*Isab* and *Sadler*, unpublished.)

Sadler, unpublished). Fig. 19 shows a ^{13}C NMR spectrum of an aurothiomalate-glutathione mixture in molar ratio 1 : 1. Only half of the coordinated thiomalate is displaced from gold by the glutathione and becomes free in solution as thiomalate-SH at this pH. We are doing further work to determine the involvement of 2 : 1 complexes of the type R′S−Au−SR or R′S(H)−Au−SR, and molecular clusters, and the mode of breakdown of the polymeric AuSR species by added thiols.

The observation by *Gerber* (*61*) that thiomalate is not released when aurothiomalate binds to albumin may also be a result of the formation of such 2 : 1 thiol Au^I complexes, if the Cys residue of albumin is involved in binding.

Another consequence of fast ligand exchange on Au^I may be the reported failure of attempts (*87, 88*) to resolve dissymmetric, tetrahedral Au^I salts such as:

197

4.7. Aurothioglucose and Obesity

The injection of aurothioglucose into mice results in hyperphagia[6]) and obesity. There is a rapid deposition of tissue lipid, and increases in body weight of 33% are common. Au^I thiomalate, thiocaproic acid, thioglycerol, thioglycoanilide and thiosulphate are inactive (89), and it is clear that the glucose component increases the ability of this compound to permeate the ventromedial hypothalmic barrier. Changes in liver and adipose tissue enzyme have been noted (90), but little is known about the fate of the gold ion.

4.8. Au^I Phosphine Drugs

Aurothiomalate, and probably many of the other Au^I thiols mentioned so far are ineffective as anti-inflammatory drugs when administered *orally*. Et_3PAuCl, on the other hand, is recently reported (91, 92) to be as effective orally as aurothiomalate given parenterally (intramuscularly), at least in suppressing the inflammatory lesions of adjuvant-induced *(Mycobacterium butyricum)* arthritis in rats. Moreover, there is much less retention of gold in the kidney. For the series of compounds:

$$R_3PAuCl, \quad R = Me, \quad Et, \quad i-Pr, \quad n-Bu$$

R = Et gives the highest therapeutic effect and serum Au level (5.38 μg Au/ml serum after dosage of 10 mg/kg/day calculated on Au) (93). Activity is also observed when R_3P is a phosphite.

Phosphine ligands themselves gave no protection against adjuvant arthritis. The chloride ligand is easily replaced by a thiol:

$$R_3PAuCl + R'S^- \rightleftarrows P_3PAuSR' + Cl^-$$

When R = Et and R' = Me, Et, Pr, Bu, Bu^i, C_6H_5, $C_6H_5CH_2$ or $\beta-C_{10}H_7$ the mixed phosphine-thiol is soluble in chloroform but unstable (94). PEt_3 is slowly liberated. Other mixed complexes are reported to have oral, anti-inflammatory activity: when thiol = thioethanol, thioglucose, thioimalate, thiobenzoic acid or thiophenol (93, 95). However, activity could be due to the formation of R_3PAuCl *in vivo* (reverse of the above reaction), since the thiol is liberated as R'SH in acid halide solutions (94).

The compounds R_3PAuCl are usually soluble in organic solvents and not in water, and might be expected to interact with lipid media, such as membranes, *in vivo*. However, in view of the many possibilities for chemical change, via combination with cysteine or protein thiols for example, this may not be the active species. The ready

---

[6]) hyperphagia: increased macrophage level.

release of gold thiols in the presence of the phosphine ligand may explain the easy passage through the kidney, which appears to have many thiol traps for heavy metals (e.g. metallothionein (96)).

The dimeric gold complex I also shows oral absorption properties (92),

$$Et_2PCH_2-CH_2-S$$
$$\qquad Au \qquad\qquad Au$$
$$\qquad S-CH_2-CH_2-PEt_2 \cdot$$

$$(C_6H_5)_2PCH_2CH_2P(C_6H_5)_2$$
$$\qquad Au \qquad\quad Au$$
$$(- \qquad\qquad SCH_2CH_2S-)_n$$

$$1 \qquad\qquad\qquad\qquad\qquad 2$$

although the polymer 2 is ineffective. Dimeric Au^I dithiocarbamates are also ineffective orally.

A comparison of the mechanism of action of these drugs with the water soluble Au^I thiols awaits information on excretion (is the phosphine still bound to Au?) and macrophage uptake patterns.

5. Au^{III}: Auric Compounds

Auric compounds have often been used in biological studies with the assumption that their behaviour would resemble that of aurous compounds. For example, neutral solutions of auric chloride were advocated in 1906 (97) as non-injurious eye washes with powerful antiseptic action.

However, many Au^{III} compounds are strong oxidising agents, and if used as drugs could be reduced to Au^{I} *in vivo*. (Tetrasuccinimido Au^{III})$^-$ is effective (2) against pleuropneumonia-like organisms in mice, although we see from Fig. 1 that N ligands confer some stability on Au^{III}. Thus we also consider substitution reactions of Au^{III} complexes.

5.1. Oxidising Action

The reaction between AuCl$_4^-$ and methionine derivatives illustrates the oxidising properties of Au^{III}. In Fig. 20 ^{1}H NMR is used to follow the course of such a reaction (98). Methionine is oxidised the sulphoxide (this is said to be stereospecific (99)), and the Au^{I} produced coordinates with excess methionine via the S atom of the thioether linkage. On standing, or at Met: Au^{III} ratios less than 2 : 1 metallic Au is deposited. Fast exchange of free and coordinated Met at Au^{I} is apparent. Substitutions at Au^{III} usually take place more slowly.

5.2. Ligand Exchange and Substitution on Au^{III}

Because of the ease of reduction of Au^{III} in aqueous solution to Au^{I} and eventually metallic Au, very little is known about the mechanism of substitution reactions in this medium. In non-aqueous media such as methanol or acetone, reactions of square-planar Au^{III} complexes tend to occur via 5-coordinate intermediates, and Au^{III} discriminates in favour of the more polarisable ligands, Table 8, that is a reactivity order:

$$Cl^- < Br^- < I^- .$$

This discrimination is much greater than observed for the isoelectronic Pt^{II} ion (100).

Fig. 20. 100 MHz ^{1}H NMR spectra of N-acetyl-L-methionine in ^{2}H$_2$O in the presence of varying ▷ amounts of AuCl$_4^-$ (molar equivalents). The S−Me peak gradually moves downfield (coordination to Au^{I}) and diminishes in intensity, while peaks from the product, N-acetyl-L-methionine sulphoxide, increase in intensity. Oxidation is complete at a 1:1 molar ratio (98).

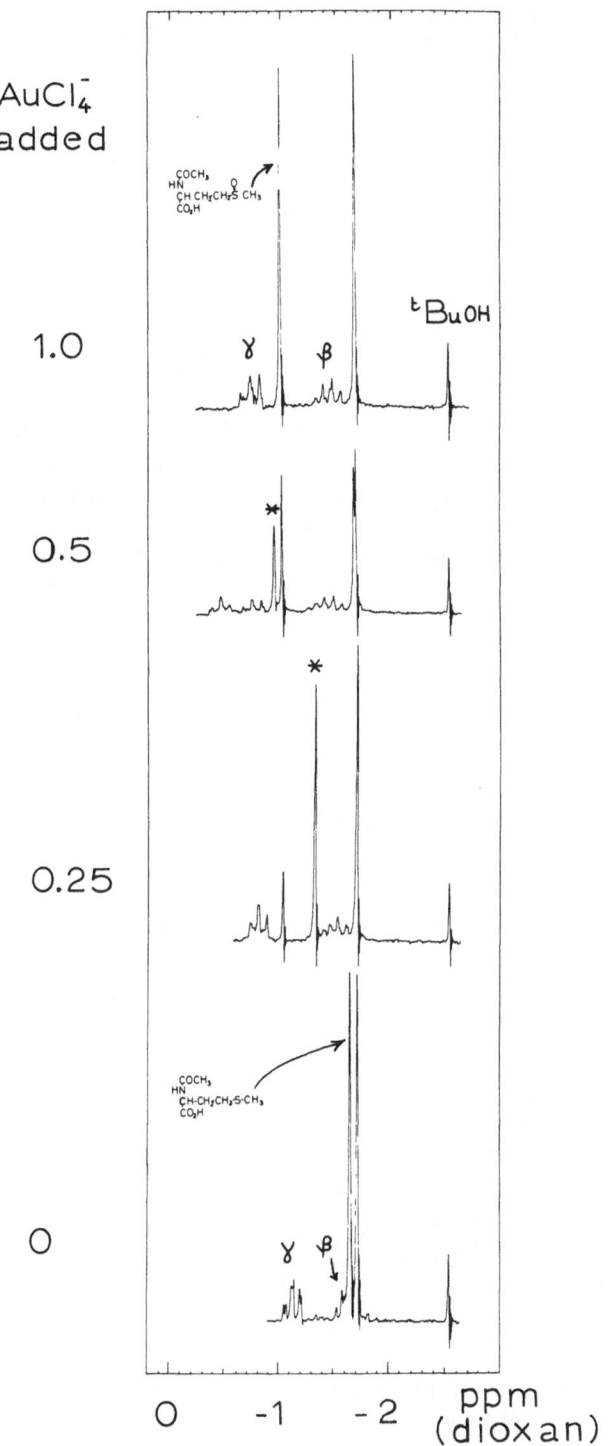

Table 8. Rate constants for the displacement reactions
$MCl_4^- + X^- \rightarrow MCl_3X^- + Cl^-$ in MeOH (Au^{III}) or H$_2$O(Pt^{II}),
25 °C (from [100]).

Entering group	Specific rate constant, K (M^{-1} sec^{-1})	
X	M = Au	M = Pt
NO$_2^-$	0.028	0.0009
Br$^-$	0.140	0.0018
I$^-$	84	0.0065
SCN$^-$	7.4	0.0021

The rate of hydrolysis of AuCl$_4^-$ in water is 375 times greater than that of PtCl$_4^{2-}$. As shown in Fig. 21 this is a stepwise process (101). The first and second hydrolyses take about 1 minute and 1 hour respectively, but the third and fourth take several hours.

The slowness of exchange on Au^{III} when compared to Au^I is apparent. For example, addition of free phosphine ligand to (CH$_3$)$_3$AuPMe$_3$ causes no change in the NMR spectrum of bound phosphine even up to 60 °C, compare Au^I in Fig. 18 (84). Similarly, although the inversion at sulphur is fast in the thioether Au^I complex (PhCH$_2$)$_2$SAu^{I}Cl, there is no inversion even at very low temperatures in (PhCH$_2$)$_2$ SAu^{III}Cl$_3$ (102). Both examples indicating the increased strength of an Au^{III}–X bond compared to Au^I–X.

It seems unlikely that substitution and exchange reactions proceed via 3-coordinate Au^{III}. Me$_2$AuOH, for example, does not show 3 coordination but polymerises to

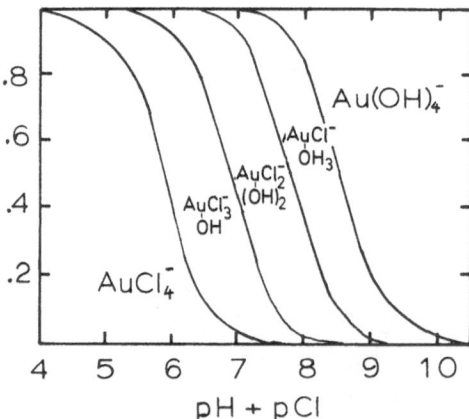

Fig. 21. Dependence of the distribution of hydrolysed AuCl$_4^-$ complexes on the concentrations of H$^+$ and Cl$^-$ ions (20 °C, I = 0.01 from (101)).

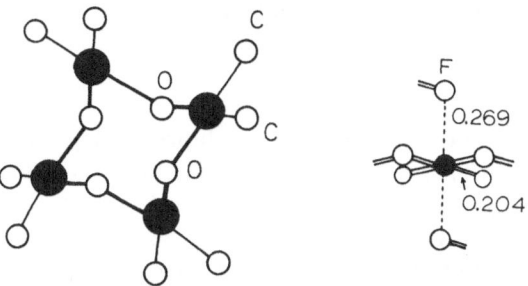

Fig. 22.
(A)   Me$_2$AuOH achieves a square-planar coordination sphere for the Au(III) atom by forming a tetramer. The eight-membered ring is puckered, and the methyl groups are non-equivalent in the ^{1}H NMR spectrum (103).
(B)   AuF$_3$ also does not contain 3-coordinate Au(III), but is polymeric. Each square-planar unit is linked to two adjacent units by cis fluorine bridges giving an infinite hexagonal helix. Longer Au––F bonds crosslink the chains, forming a tetragonally elongated octahedral environment for each gold ion (104).

achieve square-planar, 4-coordination (103), Fig. 22. 5- and 6-coordinate Au^{III} complexes, on the other hand, are known (105, 106), Figs. 23 and 24. Although interactions in the 5th and 6th positions seem to be particularly weak, they probably provide the key to the mechanism of reaction of four-coordinate Au^{III} complexes.

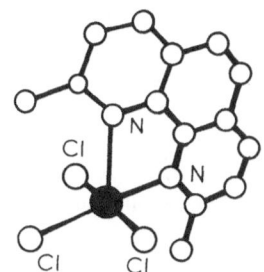

Fig. 23. The 5-coordinate complex trichloro-(2,9-dimethyl-1,10-phenanthroline) Au (III). The distribution of ligands is approximately square-pyramidal with a long axial Au–N distance (0.258 nm). Steric interactions (CH$_3$–––Cl) prevent more than one Cl sharing the dmp plane.(105).

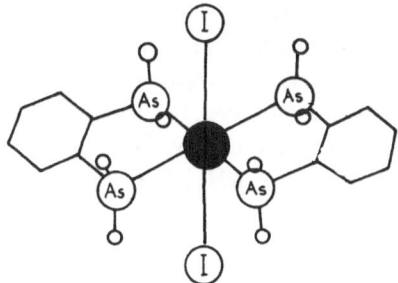

Fig. 24. The configuration of the diiodobis-(o-phenylenebis(dimethylarsine) Au(III) cation. There are four short coplanar Au–As bonds (0.243 nm) and two very long axial Au–I bonds (0.335 nm) (106).

203

5.3. Labelling Proteins and Nucleotides with Au^{III}

(i) Proteins

$AuCl_4^-$ causes aggregation of native *ribonuclease A* in solution (98), probably via displacement of chloride ligands by NH_2 groups of lysine residues on the surface of the enzyme. This method of aggregation has also been suggested for the aggregation of ovalbumin by chloroaurate (107). However, when ribonuclease is partially unfolded at pH 1.5, $AuCl_4^-$ causes oxidation of methionines to the sulphoxide, and further protein unfolding (85). Although there is one solvent-accessible methionine residue in native ribonuclease, the mechanism of oxidation by Au^{III} may require simultaneous coordinations of two methionine S atoms (99) and this may be sterically unfavourable in contrast to methionine itself or simple derivatives.

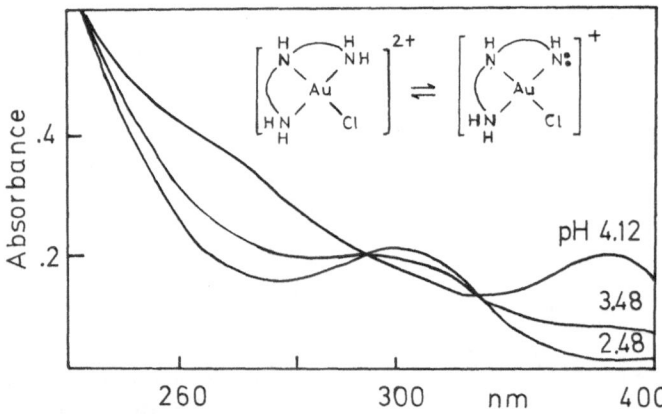

Fig. 25. Absorption spectra of 2.5×10^{-4} M [Au(dien)Cl]Cl$_2$. Loss of a proton occurs from a coordinated NH_2 group (pK$_a$ = 4) and the colourless solution becomes yellow as the pH is raised from 2.48 to 4.12 (111).

The high polarising power of Au^{III} is evident in its interactions with NH_2 groups. A coordinated NH_2 group in Au(ethylenediamine)$_2^{3+}$ readily loses a proton with pK$_a$ = 6.5 (108, 109). In the free ligand this pK value is difficult to measure, but is about 18.2 (110). A colour change of the complex ion from colourless to yellow occurs when a proton is lost from Au(en)$_2^{3+}$, and a similar change is observed (111) in diethylenetriamine Au^{III} complexes, see Figure 25. Similarly, Au(NH$_3$)$_4^{3+}$ is a weak acid,

$$Au(NH_3)_4^{3+} \rightleftharpoons Au(NH_3)_3NH_2^{2+} + H^+$$

with $pK_a = 7.48$, lM sodium perchlorate, 25 °C (13). The amido complex readily hydrolyses in water to give $Au(NH_3)_3OH^{2+}$, the well-known "fulminating gold" which is explosive when dry.

There are several reported attempts to label crystalline proteins with Au^{III} in order to help phasing problems in X-ray structure determinations, some are listed in Table 9. At pH 4.5 in the presence of excess Cl^-, $AuCl_4^-$ binds as an anion to *lysozyme* between two positively charged arginine residues (112). A curiously-slow reaction with sperm-whale *myoglobin* has been observed (113, 114). In the ammonia buffer at pH 7 Au^{III} ammonia complexes such as $Au(NH_3)_4^{3+}$ may be formed, presumably together with $Au(NH_3)_3NH_2^{2+}$. The latter, with its coordinated NH_2^-, group ought to be a reactive species, especially toward carbonyl groups. The gold ion was eventually located in the structure in the same site as Ag^+ in another derivative. Perhaps during the long reaction time Au^{III} is reduced to Au^I. Met 55 would be accessible to $AuCl_4^-$, but Met 131 is buried.

Cys SH groups on proteins may provide strong binding sites for Au(III), but oxidation can again occur. A common route to Au(I) mercaptides involves reaction of Au(III) with 3 moles of the mercaptide, the disulphide being the other product.

Table 9. Au^{III} binding sites on crystalline proteins.

Protein	Conditions	Au Site[a]	Reference
Lysozyme	$AuCl_4^-$, 3% lysozyme 5% NaCl, pH 4.5	between Arg 14 of two symmetry related molecules (same site as HgI_3^-)	112
Myoglobin (seal, or sperm whale)	$NaAuCl_4$, NaCl, 4 M $(NH_4)_2SO_4$, pH 6–7 (Crystals grow very slowly, 6–9 months)	between His B5 and His GH1 (same as Ag^+)	113, 114
	$KAuI_4$	close to haem (same as HgI_3^-)	
Lactate dehydrogenase	1 mM $NaAuCl_4$	2 sites, one at Cys?	115
Yeast phosphoglycerate kinase	20 mM $NaAuCl_4$ $(NH_4)_2SO_4$	at surface (Cys ?)	116
α-Chymotrypsin	1 mM $KAuI_4$ $(NH_4)_2SO_4$, pH 4	?	117
Bence-Jones protein (immunoglobulin constant region, light chain dimers)	5 mM $KAuI_4$ 2 M $(NH_4)_2SO_4$ pH 4.5, phosphate-citrate buffer	?	118

[a] The resolution obtained in X-ray structure determinations is usually not sufficient for a full definition of the coordination sphere of protein bound Au.

The problem of AuI_4^-

There are several reports of protein labelling with AuI_4^-, Table 9. However, we should not expect this to be the form of the complex which binds. Because of the high polarising power of Au^{III}, AuI_4^- spontaneously disproportionates in aqueous solution (*108, 119, 120*):

$$AuI_4^- \rightarrow AuI_2^- + I_2 \ (\rightarrow IAu(I_3)^- ?)$$

$KAuI_4$ exists as a black solid, but is thermally unstable, liberating I_2. The disproportionation of AuI_4^- has been confirmed spectroscopically (*108*). The liberation of I_2 is demonstrated in Fig. 26.

Little is known about the stability or reactivity of AuI_2^- in aqueous solution. In acetonitrile it is a pale yellow ion, see Figure 27, and very stable ($\log k_1 k_2 = 23.8$) (*9*). However it appears that $[Bu_4^nN][AuI_2]$ immediately deposits metallic Au when H_2O is added to ethanolic solutions (*121*).

(ii) Nucleotides

Wulff (*122*) in 1893 isolated insoluble 1:1, Au: adenine adducts from a reaction of $AuCl_4^-$ with adenine in dilute HCl. Subsequently, 2 Au: 1 adenine adducts were isolated by a modified procedure (*123*). These were also insoluble in water, as were 1:1 adducts with guanine, xanthine and hypoxanthine (*124*). Mixtures of $AuCl_4^-$ and adenine nucleotides change colour from yellow to dark brown in 1 hour (*125*), and provided the only cations present are Na^+ or K^+ the adducts remain soluble in aqueous

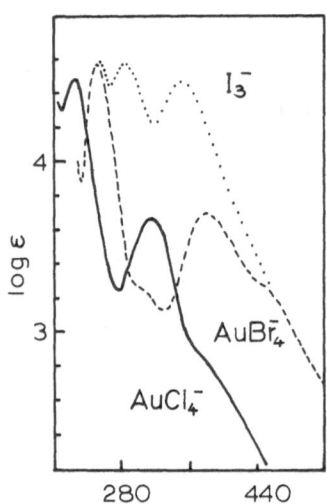

Fig. 26. Electronic absorption spectra of $KAuCl_4$ in 1 N KCl (———), $KAuBr_4 \cdot 2 H_2O$ in 1 N KBr (———), and $KAuCl_4$ or $KAuBr_4 \cdot 2 H_2O$ in 1 N KI (········). AuI_4^- disproportionates and I_3^- is liberated (*108*).

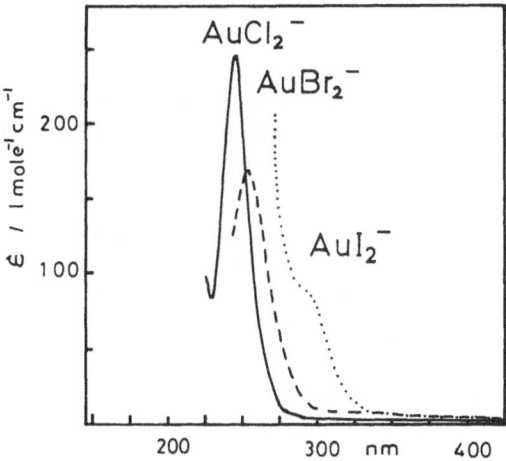

Fig. 27. Electronic spectra of the colourless ions $\overset{I}{Au}Cl_2^-$ and $\overset{I}{Au}Br_2^-$, and pale yellow ion AuI_2^- in acetonitrile at 20 °C, prepared by mixing in 1:3 mole ratios of $Au(MeCN)_2ClO_4$ with Et_4NCl and Bu_4NBr, and $AuBr_2^-$ with excess Bu_4NI (9).

media. By ethanol precipitation, brown water-soluble products are obtained. The nucleotide phosphate groups solubilize these compounds, but the sugar moiety does not reduce the Au^{III} ion to Au^0, as does ribose, deoxy-ribose 5-phosphate, sucrose or dextrose. Little is known about the structure of the water soluble adducts, but they may be polymeric with coordination to NH_2, N–6 and N–7 of adenosine. They have potential applications in cytochemistry, staining chromatin, nucleoli and ribosomes, but not cell membranes. Some of the Au is reduced to Au^0 by tissue components.

Since Au^{III} adenosine monophosphate is a soluble complex, but Au^{III}-adenosine is insoluble, it is possible to demonstrate by electron microscopy the localisation of acid phosphatase in lysosomes (126). The phosphate group is cleaved by the enzyme and an electron-dense material deposited at the enzyme site.

A brown precipitate is obtained (127) by ethanol precipitation from chloroaurate/ DNA mixtures. Time dependent viscosity changes are observed in the mixture, but are reversed by CN^-. Gold binding to base and phosphate on DNA was suggested.

Polyadenylic acid can be stained with $HAuCl_4$ (pH 4.5) to give two gold atoms per nucleotide (128). This produces a significant improvement in the optical density of the image viewed by electron microscopy. Polyuridylic acid (no NH_2 groups) on the other hand is not stained by $HAuCl_4$.

A recent report (129) that the 5-diazouracil (5–du) complex of Au^{III}, $[Au(5-du)_2Cl_2]Cl \cdot HCl$, shows anti-tumour activity in mice, introduces a new possible pharmacological role for Au^{III}. Cis-dichlorodiamine Pt^{II} complexes (isoelectronic with Au^{III}) are already known which show anti-tumour activity (130), but an increased reactivity is to be expected with Au^{III}.

6. EPR detection of Au^{II}

Although Au^{II} compounds are very unstable in aqueous solution, they may be of importance as intermediates in redox reactions of Au^{I} and Au^{III}, when one-electron redox-reagents are used (the reduction of AuCl$_4^-$ by PPh$_3$ on the other hand, presumably involves a 1-step, 2-electron transfer process, and would not use Au^{II} as an intermediate). Several stable compounds were once thought to contain Au^{II}, but actually contain Au^{I}/Au^{III} mixtures: (C$_6$H$_5$CH$_2$)$_2$SAuCl$_2$ (*131*) CsAuCl$_3$ (*132*), Au(dimethylglyoxime)Cl (*79*), the caesium salt being Cs$_2$[(AuCl$_2$)$^-$ (AuCl$_4$)$^-$]. A few authentic Au^{II} compounds are known, Table 10, and these can usually be recognised by a characteristic 4-line EPR spectrum at room temperature, Fig. 28. Au^{I} dialkyldithiocarbamates are oxidised by the corresponding thiuram disulphides in benzene solution to produce Au^{II} dialkyldithiocarbamates (*134*). Cystine disulphides are available in many proteins as possible 1-electron oxidising agents.

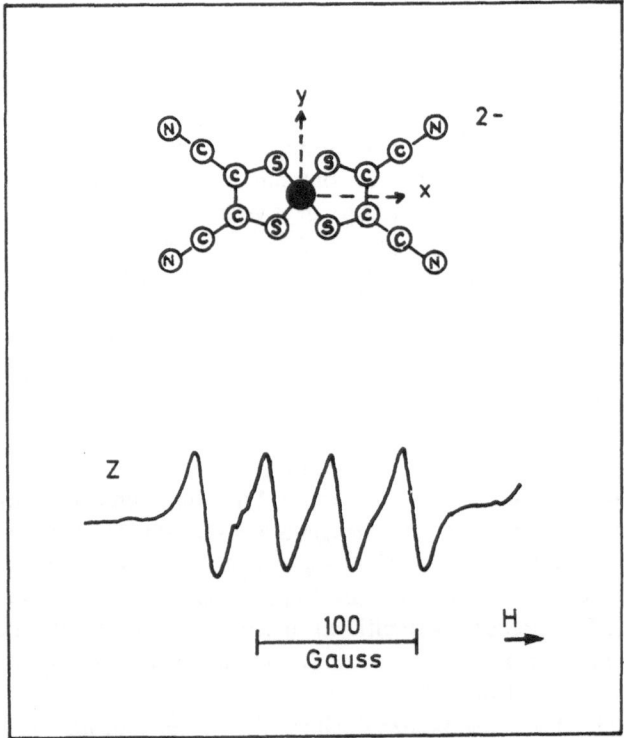

Fig. 28. Electron paramagnetic resonance spectrum of a single crystal containing Au^{II}[S$_4$C$_4$(CN)$_4$]$^{2-}$ at room temperature (*135*). The 4 line pattern is typical of coupling to ^{197}Au, I = 3/2, and has also been observed for Au^{II} N,N-dialkylthiocarbamates in benzene solution.

Table 10. Au^{II} Compounds

Compound	Reference
1. Au^{II} phthalocyanine	133
2. bis(diethyldithiocarbamato) Au^{II}	134
3. bis(maleonitriledithiolato) Au^{II}	135, 136
4. π-(3)-1,2-dicarbollyl Au^{II}	137

7. Mössbauer confirmation of Au^V

It is unlikely that Au^V will have any biological importance, but it is worth noting the Mössbauer results on Au^V compounds (*138*) and comparing them with other oxidation states. Au^V is stabilised by F$^-$ ligands in octahedral complexes such as Cs$^+$AuF$_6^-$. These are extremely moisture sensitive. The isomer shifts and quadrupole couplings are shown in Fig. 29 and compared to those of Au^0, Au^I and Au^{II}. The large difference in quadrupole splitting between Au^I and Au^{III} is apparent, but there is no satisfactory explanation at present (*139*).

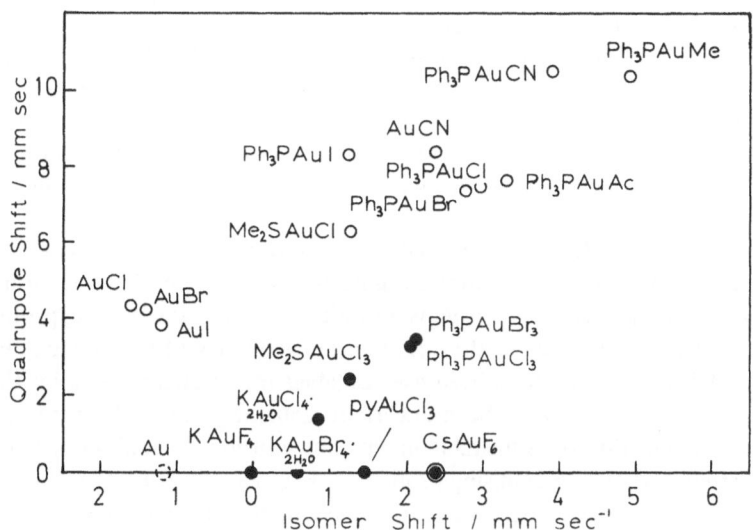

Fig. 29. Mössbauer parameters for Au (o), Au^I (o), Au^{III} (•) and Au^V (o) compounds at 4.2 °K relative to ^{197}Pt source Data from (*138, 139*).

8. Conclusion

In this article we have attempted to seek a molecular basis for the biological effects of gold compounds. Such an understanding is important in the two major areas of interest in the biological chemistry of gold: as a metallo-drug and heavy-atom label. In both applications *specificity* of action is required of the gold compound, and emphasis has been placed on knowledge of thermodynamics and kinetics of gold-binding to small and large ligands, and, in particular, on the coordination geometry at the gold atom.

Au(I) has anti-inflammatory activity, but, although $Au(CN)_2^-$ is one of the most stable Au(I) ions in solution, it is too toxic for clinical use. The simple Au^+ cation does not appear to exist in aqueous solution and most Au(I) compounds are insoluble in water and unstable in its presence (*140*). Mercaptides stabilise Au(I) in water, and aurothiomalate is in widespread clinical use as an anti-inflammatory drug. The nature of the active species of this drug is unclear. *In vivo* scrambling amongst available sulphydryl binding sites is probable. The gold ion seems to enter many cells, but becomes localised within the lysosomes of the phagocytic cells called macrophages. Here they may inhibit enzymes important in inflammation. Au(I) may discriminate between SH groups within the pool. Some will be sterically inaccessible (using a well-established example as illustration: haemoglobin has three cysteine SH groups: α 104, β 112, β 93, but only β 93 is freely titratable by thiol reagents) and their states of protonation may vary (whereas the pK_a values of the SH groups of cysteine itself, 10.5, and glutathione, 9.7, are high, values are often lowered in enzymes, perhaps as low as 5). This ease of ligand exchange on 2-coordinate Au(I) is probably accounted for by its ease of formation of 3-coordinate intermediates.

In a study of natural anti-tumor mechanisms in animals it has been found (*141*) that prior injection of aurothiomalate enhances the "take" of a subcutaneous tumour innoculum, which suggests that anti-tumour mechanisms are macrophage mediated. It remains to be seen whether Au(I) phosphine chlorides also become localised in macrophages. Although they are not water soluble, they can be administered orally as anti-inflammatory agents.

Knowledge of Au(I) binding sites on large molecules such as proteins is unfortunately limited to a few studies using $Au(CN)_2^-$. There is a need for more gold labelling studies on crystalline proteins, not just to solve X-ray phasing problems, but for the specific investigation of the coordination geometry of the bound gold ion. When Au(III) compounds have been used as labels for crystalline proteins the nature of the bound species has often been uncertain. Labelling with AuI_4^-, for example, has been claimed, but this ion is unstable in aqueous solution. In addition, Au(III) compounds often have strong oxidising properties. With a careful choice of ligands for Au(III), a new range of anti-tumour drugs may emerge, for the ion is known to have a high affinity for polynucleotides, and hence may interfere with cell division processes.

The use of colloidal gold for labelling macrophages, proteins and immunoglobulins presents some intriguing problems for the chemist. A clearer definition of the gold atom on the surface is required. The labelling may be non-specific, but, since labelled anti-bodies often retain full biological activity, stronger directional forces may be involved, perhaps through Au(I) on the surface.

It is now becoming clear that the biochemical behaviours of heavy metal ions, although showing some similarities, particularly in their affinity for polarisable ligands, will show important differences. Some of these have already been noted (*142*). For example, Hg^{2+} and Cd^{2+} are much more powerful inhibitors of enzymes bearing a single sulphydryl functional group than Pb^{2+}, Hg^{2+} has little affinity for phospholipids but Pb^{2+} binds firmly to phosphatidylcholine membranes in vitro. Although the data is limited, Au^+ appears to have a much lower affinity for amino and carboxylate groups than Hg^{2+}. Neither Hg^{2+} nor Ag^+, both d^{10} ions, have available stable higher oxidation states in aqueous media. These considerations point to a *unique biochemical behaviour for gold.*

In the field of chemotherapy gold drugs offer potential advantages over organic substances in mechanism of action studies by their probe properties. They should be viewed not only as a potential cure for a disease but also as a probe for discovering the cause.

Note added in proof

Dr. S. Lindskog has informed me that the residue labelled His 128 in Fig. 8 (taken from ref. 46) is now known to be Phe 129, which cannot hydrogen bond to $Au(CN)_2^-$. The position of the gold atom was clear in the electron density maps at that time, but not the details of the gold coordination sphere. We await further refinement to see if an increase in coordination number has occurred, or indeed if either of the cyanide ligands has been displaced.

Acknowledgements. I am very grateful to *Dr. B. Vernon Roberts* and *Mr. J. L. Doré* (The London Hospital Medical College), *Dr. C. Danpure, Dr. A. M. Denman* and *Dr. J. M. Gumpel* (Clinical Research Centre, Northwick Park) and *Dr. P. D'Arcy Hart* (National Institute for Medical Research, Mill Hill) for discussion and supplying results and material as indicated in the text.

P. J. Sadler

References

1. *Dyson, G. M.:* Pharm. J. *123*, 249–50 and 266–7 (1929).
2. *Block, W. D., Van Goor, K.:* Metabolism, Pharmacology and Therapeutic Uses of Gold Compounds. American lect. Ser. 282. Springfield, Illinois: Thomas, C. C. 1954.
3. *Koch, R.:* Deutsche med. Wochenschr. *16*, 756 (1890).
4. *Forestier, J.:* Lancet *2*, 441 and 646 (1932).
5. *D'Arcy Hart, P.:* Brit. Med. J. Now. 30 and Dec. 7, 805 and 849 (1946).
6. *Landé, K.:* Münch. Med. Wochenschr. *74*, 1132 (1927).
7. *Constable, T. J., Crockson, A. P., Crockson, R. A., McConkey, B.:* Lancet, May 24, 1176–1179 (1975).
8. *Phillips, C. S. G., Williams, R. J. P.:* Inorganic Chemistry, Volume 2, O. U. P., (1966).
9. *Roulet, R., Lan, N. Q., Mason, W. R., Fenske, G. P.:* Helv. Chim. Acta *56*, 2405–18 (1973).
10. *Goolsby, A. D., Sawyer, D. T.:* Anol. Chem. *40*, 1978–83 (1968).
11. *Hawkins, C. J., Moensted, O., Bjerrum, J.:* Acta. Chem. Scand. *24*, 1059–66 (1970).
12. *Ford-Smith, M. H., Habeeb, J. J., Rawsthorne, J. H.:* J. Chem. Soc. Dalton, 2116–2120 (1972).
13. *Skibsted, L. H., Bjerrum, J.:* Acta Chem. Scand. (A) *28*, 740 (1974).
14. *Pouradier, J., Gadet, M. C.:* J. Chim. Phys. Physiochim. Biol. *66*, 109–112 (1969).
15. *Pouradier, J., Gadet, M. C.:* J. Chim. Phys. *62*, 1181–6 (1965).
16. *Pouradier, J., Gadet, M. C.:* J. Chim. Phys. *63*, 1467–73 (1966).
17. *Feuerstein, H., Oschinski, J.:* Z. Naturforsch. Teil C. *29*, 414–6 (1974).
18. *Petros, H., McMillan, A. L.:* Brit. J. Dermatol. *88*, 505–508 (1973).
19. *Talipov, R. M., Khatamov, Sh.:* Uzb. Geol. Zh. *17*, 26–31 (1973).
20. *Talipov, R. M., Khatamov, Sh.:* Uzb. Geol. Zh. *18*, 23–7 (1974).
21. *Korobushkina, E. D., Chernyak, A. S., Mineev, G. G.:* Mikrobiologiya, *43*, 49–54 (1974).
22. Colloid Chemistry, 2nd Edition, Oxford: Pergamon Press. Ed. by *Jirgensons, B., Straumanis, M. E.*
23. *Frens, G.:* Nature. Phys. Sci. *241*, 20–22 (1973).
24. *Gosselin, R. E.:* J. Gen. Physiol. *39*, 625–649 (1956).
25. *Cohn, Z. A., Benson, B.:* J. Exp. Med. *122*, 455 (1965).
26. *Hirsch, J. G., Fedorko, M. E., Cohn, Z. A.:* J. Cell Biol., *38*, 629–632 (1968).
27. *Pauli, W.:* Trans. Farad. Soc. 31, (1935). Helv. Chim. Acta. *32*, 795 (1949).
28. *Meade, C. J., Lackman, P. J., Brenner, S.:* Journal of Immunology, *27*, 227 (1974).
29. *Faulk, W. P., Taylor, G. M.:* Immunochemistry, *8*, 1081–1083 (1971).
30. *Romano, E. L., Stolinski, C., Hughes-Jones, N. C.:* Immunochemistry, *11*, 521–2 (1974).
31. *Horisberger, M., Rosset, J., Bauer, H.:* Experientia, *31*, 1147–49 (1975).
32. *Sabin, A. B., Warren, J.:* Bacteriol. *40*, 823–56 (1940).
33. *ERC trial,* Ann Rheum. Dis.: *20*, 315 (1961).
34. *British Medical Journal,* April 26, 156–7 (1975).
35. *Courmont, P., Gardère, H., Pichat, P.:* Compt. rend. soc. biol. *112*, 892–4 (1933).
36. *Novick, R. P., Roth, C.:* J. Bacteriol. *95*, 1335–42 (1968).
37. *Moore, B.:* Lancet, *2*, 453–8 (1960).
38. *Davidson, M., Thomas, L.:* Antimicrob. Agents and Chemother., 312–315 (1966).
39. *Ruben, H., Zalkin, A., Faltens, M. O., Templeton, D. H.:* Inorg. Chem. *13*, 1836–39 (1974).
40. *Norton, W. L., Lewis, D. C., Ziff, M.:* Arth and Rheum, *11*, 437– (1968).
41. *Vernon-Roberts, B., Jessop, J. D., Doré, J.:* Ann. Rheum. Dis. *32*, 301, (1973).
42. *Jessop, J. D., Vernon-Roberts, B., Harris, J.:* Arth. and Rheum. *32*, 294 (1973).
43. *Persellin, R. H., Ziff, M.:* Arth. Rheum. *9*, 57 (1966).
44. *Söderberg, B.-O., Zeppezauer, E., Boive, T., Nordström, B., Bräden, C.-I.:* Acta. Chem. Scand. *24*, 3567–74 (1970).
45. *Gunnarsson, P.-O., Pottersson, G.:* Eur. J. Biochem.: *43*, 479–86 (1974).
46. *Lindskog, S., Henderson, L. E., Kannan, K. K., Liljas, A., Nyman, P. O., Strandberg, B.:* The Enzymes, Ed. by *P. D. Boger.* Vol. V. p. 621. (1971).

47. *Kannan, K. K., Norstrand, B., Fridborg, K., Lörgren, S., Ohlsson, A., Petef, M.:* Proc. Natl. Acad. Sci. USA, *72*, 51 (1975).
48. *Dolhie, H.-J., Llewellyn, F.-J., Wardlaw, W., Welch, A. J. E.:* J. Amer. Chem. Soc., *61*, 426 (1939).
49. *Chadwick, B. M., Sharp, A. G.:* Adv. Inorg. Chem., *8*, 83 (1966).
50. *Westwick, W. J., Allsop, J., Watts, R. W. E.:* Biochem. Pharm. *23*, 153–162 (1974).
51. *Westwick, W. J., Allsop, J., Watts, R. W. E.:* Biochem. Pharmac. *23*, 163–165 (1974).
52. *Janoff, A.:* Biochem. Pharmac. *19*, 626–8 (1970).
53. *Koelle, G. B.:* J. Histochem. Cytochem. *22*, 252–9 (1974).
54. *Lorber, A., Bovy, R. A., Chang, C. C.:* Nature New Biol. *236*, 250–52 (1972).
55. *Mascarenhas, B. R., Granda, J. L., Freyberg, R. H.:* Arth. and Rheum. *15*, 391–402 (1972).
56. *Gerber, R. C., Paulus, H. E., Bluestone, R., Lederer, M.:* Arth. and Rheum., *15*, 625–629 (1972).
57. *Gerber, D. A.:* J. Pharm. Exp. Ther. *143*, 137–140 (1964).
58. *Danpure, C. J.:* Biochem. Soc. Trans., 549th meeting, 899–901 (1975).
59. *Hull, H. H., Chang, R., Kaplan, L. J.:* Biochim. Biophys. Acta, *400*, 132–36 (1975).
60. *Britten, M. C., Schur, P. H.:* Arth. Rheum. *14*, 87 (1971).
61. *Gerber, D. A.:* Arth. and Rheum. *17*, 85 (1974).
62. *Denman, E. J., Denman, A. M.:* Ann. Rheum. Dis. *27*, 582 (1968).
63. *Baker, R. W., Pauling, P. J.:* J. Chem. Soc. Dalton, 2264–66 (1972).
64. *Wilford, J. B., Powell, H. M.:* J. Chem. Soc. (A), 8–15 (1969).
65. *Baenziger, N. C., Dittmore, K. M., Royle, J. R.:* Inorg. Chem. *13*, 805–810 (1974).
66. *Isab, A. A., Sadler, P. J., Danpure, C. J., Fyfe, D. A., Charlwood, P. A.:* manuscript in preparation.
67. *Moore, E. E., Ohman, R. J., March, B.:* J. Ann. Pharm. Soc. *40*, 184–5 (1951).
68. *Shugam, E. A., Zhdanov, G. S.:* Acta Physicochim. USSR, *20*, 247–52 (1945).
69. *Janssen, E. M. W., Folmer, J. C. W., Wiegers, G. A.:* J. Less Comm. Metals, *38*, 71–76 (1974).
70. *Hong, S.-H., Olin, A., Hesse, R.:* Acta Chem. Scand. *A 29*, 583–589 (1975).
71. *Hesse, R., Nilson, L.:* Acta Chem. Scand. *23*, 825–845, (1969).
72. *McPartlin, M., Mason, R., Malatesta, L.:* Chem. Comm. 334 (1969).
73. *Hesse, R., Jennische, P.:* Acta. Chem. Scand., *26*, 3855–64 (1972).
74. *Batchelder, D. N., Simmons, R. O.:* J. Appl. Phys. *36*, 2864 (1965).
75. *Lawton, S. L., Rohrbaugh, W. J., Kokotailo, G. T.:* Inorg. Chem. *11*, 2227–2233 (1972).
76. *Albano, V. G., Bellon, P. L., Manassero, M., Sansoni, M.:* Chem. Comm. 1210 (1970).
77. *Arai, G. J.:* Rec. Trav. Chim. *81*, 307 (1962).
78. *Drew, M. G. B., Riedl, M.-J.:* J. Chem. Soc. Dalton, 52–55 (1973).
79. *Rundle, R. E.:* J. Amer. Chem. Soc. *16*, 3101–2 (1954).
80. *Corfield, P. W. R., Shearer, H. M. M.:* Acta. Cryst. *23*, 156 (1967).
81. *Farrell, F. J., Spiro, T. G.:* Inorg. Chem. *10*, 1606–1610 (1971).
82. *Wijnhoven, J. G., Bosman, W. P. J. H., Buerskens, P. T.:* J. Cryst. Mol. Str. *2*, 7–15 (1972).
83. *Marcotrigiano, G., Battistazzi, R., Peyronel, G.:* Inorg. Nuc. Chem. Lett. *8*, 399–402 (1972).
84. *Shiotani, A., Klein, H.-F., Schmidbaur, H.:* J. Amer. Chem. Soc.: *93*, 1555–57 (1971).
85. *Spofford, W. A., Amma, E. L.:* Chem. Comm., 405–7 (1968).
86. *Vizzini, E. A., Amma, E. L.:* J. Amer. Chem. Soc., *88*, 2872–3 (1966).
87. *Davis, M., Mann, F. G.:* J. Chem. Soc. 3791–98 (1964).
88. *Cochran, W., Hart, F. A., Mann, F. G.:* J. Chem. Soc. 2816–18 (1957).
89. *Mayer, J., Marshall, N. B.:* Nature *178*, 1399–40 (1956).
90. *Martin, R. J., Lamprey, P.:* Life Sci. *14*, 1121–31 (1974).
91. *Walz, D. T., DiMartino, M. J., Sutton, B. M., Misher, A.:* J. Pharm. Exp. Ther. *181*, 292 (1972).
92. *Weinstock, J., Sutton, B. M., Kuo, G. Y., Walz, D. T., DiMartino, M. J.:* J. Med. Chem. *17*, 139 (1974).
93. *Sutton, B. M., McGusty, E., Walz, D. T., DiMartino, M. J.:* J. Med. Chem. *15*, 1095 (1972).
94. *Coates, G. E., Kowala, C., Swan, J. M.:* Aust. J. Chem., *19*, 539–45 (1966).
95. *McGusty, E. R., Sutton, B. M.:* United States Patent 3,784,687 Jan. 8 (1974).

96. *Kägi, J. H. R., Himmelhoch, S. R., Whanger, P. D., Bethune, J. L., Vallee, B. L.:* J. Biol. Chem. *249*, 3537–3542 (1974).
97. Toxicology of the Eye, 2nd Ed., Ed. by *Morton Grant* (1974).
98. *Isab, A. A., Sadler, P. J.:* manuscript in preparation.
99. *Bordignon, E., Cattalini, L., Natile, G., Scatturin, A.:* Chem. Comm., 878–9 (1973).
100. *Cattalini, L., Orio, A., Tobe, M. L.:* J. Amer. Chem. Soc., *89*, 3130–34, (1967).
101. *Bjerrum, N.:* Bull. Soc. Chim. Belg. *57*, 432–445 (1948).
102. *Coletta, F., Ettorre, R., Gambarao, A.:* Inorg. Nuc. Chem. Lett. *8*, 667–71, (1972).
103. *Glass, G. E., Konnert, J. H., Miles, M. G., Britton, D., Tobias, R. S.:* J. Amer. Chem. Soc. *90*, 1131 (1968).
104. *Einstein, F. W. B., Rao, P. R., Trotter, J., Bartlett, N.:* J. Chem. Soc. (A) 478 (1967).
105. *Robinson, W. T., Sinn, E.:* J. Chem. Soc. Dalton 726–31 (1975).
106. *Duckworth, V. F., Stephenson, W. C.:* Inorg. Chem. *8*, 1661–64 (1964).
107. *Craig, J. P., Garrett, A. G., Williams, H. B.:* J. Amer. Chem. Soc. *76*, 1570–75 (1954).
108. *Gangopadhayay, A. K., Chakravorty, A.:* J. Chem. Phys., *35*, 2206–09 (1961).
109. *Block, B. P., Bailar, J. C.:* J. Amer. Chem. Soc., *73*, 4722–25 (1951).
110. *Schaal, R.:* J. Chim. Phys. *52*, 784–95 (1955).
111. *Baddley, W. H., Basolo, F., Gray, H. B., Nölting, C., Poé, A. J.:* Inorg. Chem. *2*, 921–28 (1963).
112. *Blake, C. C. F.:* Adv. Prot. Chem. *28*, 39 (1968).
113. *Bluhm, M. M., Bodo, G., Dintzis, H. M., Kendrew, J. C.:* Proc. Roy, Soc. Lond.: *A 246*, 369 (1958).
114. *Scoloudi, H.:* Proc. Roy. Soc. Lond.: *A 258*, 181 (1960).
115. *Adams, M. J., Haas, D. J., Jeffery, B. A., McPherson, A., Mermall, H. L., Rossmann, M. G., Schevitz, R. W., Wonacott, A. J.:* J. Mol. Biol. *41*, 159 (1969).
116. *Wendell, P. L., Bryant, T. N., Watson, H. C.:* Nature New Biol. *240*, 134–136 (1972).
117. *Tulinsky, A., Mani, N. V., Morimoto, C. N., Vandlen, R. L.:* Acta Cryst. *A 29*, 1309–22 (1973).
118. *Wang, B. C., Yoo, C. S., Sax, M.:* J. Mol. Biol. *87*, 505–508 (1974).
119. *Kazakov, V. P., Erenburg, A. M., Peshchevitskii, B. I.:* Kinetics and Catalysis *6*, 645–7 (1965).
120. *Kazakov, V. P., Matreeva, A. I., Erenburg, A. M., Peshchevitskii, B. I.:* Russ. J. Inorg. Chem. *10*, 563–7 (1965).
121. *Braunstein, P., Clark, R. J. H.:* J. Chem. Soc. Dalton, 1845–48, (1973).
122. *Wulff, C.:* Z. physiol. Chem. *17*, 468 (1893).
123. *Holtz, H., Müller, H.:* Arch. Exp. Pathol. Ph. *105*, 30 (1924).
124. *Hoppe-Seyler, F. W., Schmidt, W.:* Z. Phys. Chem. *175*, 304 (1928).
125. *Gibson, D. W., Beer, M., Barrnett, R. J.:* Biochemistry, *10*, 3669–79 (1971).
126. *Fitzsimons, J. T. R., Gibson, D. W., Barrnett, R. J.:* J. Histochem. Cytochem. *18*, 673–4 (1970).
127. *Pillai, C. K. S., Nandi, U. S.:* Biopolymers, *12*, 1431 (1973).
128. *Highton, P. J., Beer, M.:* J. Roy. Micros, Soc. *88*, 23–28 (1968).
129. *Dragulescu, C., Heller, J., Maurer, A., Policec, S., Topcui, V., Csalci, M., Kirschner, S., Kravitz, S., Moraski, R.:* Int. Coord. Chem. Conf. *XVI*, 1.9, (1974).
130. *Thompson, A. J., Williams, R. J. P., Reslova, S.:* Structure and Bonding, *11*, (1972).
131. *Herrmann, F.:* Ber. Dtsch. Chem. Ges. *38*, 2813 (1905).
132. *Wells, H. L.:* Am. J. Sci. *3*, 417 (1933).
133. *Mac-Cragh, A., Koski, W. S.:* J. Amer. Chem. Soc., *87*, 2496 (1965).
134. *Vänngard, T., Akerström, S.:* Nature *184*, 183–184 (1959).
135. *Schlupp, R. L., Maki, A.:* Inorg. Chem. *13*, 44–51 (1974).
136. *Van Rens, J. G. M., Viegers, M. P. A., De Boer, E.:* Chem. Phys. Lett. *28*, 104–108 (1974).
137. *Warren, L. F., Hawthorne, M. F.:* J. Amer. Chem. Soc. *90*, 4823 (1968).
138. *Kaindl, G., Leary, K., Bartlett, N.:* J. Chem. Phys. *59*, 5050–54 (1972).
139. *Charlton, J. S., Nichols, D. I.:* J. Chem. Soc. A. 1484–88 (1970).
140. *Johnson, B. F. G., Davis, R.:* Comprehensive Inorganic Chemistry, Vol. 3, pp. 129–186, Pergamon Press (1974).
141. *McBridge, W. H., Tuach, S., Marmion, B. P.:* Br. J. Cancer, *32*, 558–567 (1975).
142. *Vallee, B. L., Ulmer, D. D.:* Ann. Rev. Biochem. 91–128 (1972).

Author-Index Volume 1–29

216

218

STRUCTURE AND BONDING

Editors: J.D. Dunitz, P. Hemmerich, R.H. Holm, J.A. Ibers, C.K. Jørgensen, J.B. Neilands, D. Reinen, R.J.P. Williams

Vol. 8.
73 figures. III, 196 pages. 1970

Vol. 11: Biochemistry
58 figures. III, 170 pages. 1972

Vol. 20: Biochemistry
57 figures. IV, 167 pages. 1974

Springer-Verlag
Berlin
Heidelberg
New York

Vol. 23: Biochemistry
50 figures. IV, 193 pages. 1975

S. Forsén, B. Lindman
Chlorine, Bromine and Iodine NMR

Physico-Chemical and Biological Applications. 72 figures, approx. 45 tables. Approx. 400 pages. 1976 (NMR Basic Principles and Progress, Vol. 12)

Contents: Introductory Aspects. — Relaxation in Molecules or Ions with Covalently Bonded Halogens. — Shielding Effects in Covalent Halogen Compounds. — Scalar Spin Couplings. — Relaxation of Chloride, Bromide and Iodide Ions. — Shielding of Halide Ions. — Quadrupole Splittings in Liquid Crystals. — Halide Ions in Biological Systems. — Studies of the Perchlorate Ion.

The NMR study of quadrupolar nuclei can provide unique and very valuable information on a variety of physico-chemical and biological systems. This volume contains a comprehensive account of published work using chlorine (^{35}Cl and ^{37}Cl) bromine (^{79}Br and ^{81}Br) and iodine (^{127}I) NMR and also outlines potential new areas of application. The theoretical background necessary for the interpretation of various NMR parameters — shieldings, spin-spin couplings, relaxation rates and quadrupolar splittings — is presented. The fields covered include: molecular motion in simple liquids, interactions and dynamics in electrolyte solutions and surfactant systems, orientation effects in liquid crystals. A considerable part of the volume is dedicated to various biological applications, for example related to the function of enzymes and other proteins.

S. Bernstein, J.P. Dusza, J.P. Joseph
Physical Properties of Steroid Conjugates
XI, 212 pages. 1968

Contents: Introduction and Scope. — Catalog of Conjugates: A. Empirical formula index of parent steroids. B. Catalog. — This book, the first compilation of its kind, presents in a concise and orderly manner the physical and "biophysical" data which characterize the conjugates, primarily sulfates and glucuronides, of 146 steroids.

The authors have made a thorough search of the extensive literature, both historical and current, and their book should constitute a ready-reference catalog of all known biologically important steroids.

This volume will prove invaluable to the biochemists, clinicians and chemists involved in the study of steroid metabolism, and the authors hope that it may also lead to a deeper understanding of the complexities of steroid metabolic transport and possibly even of disease processes.

Springer-Verlag
Berlin
Heidelberg
New York